Net-Zero and Nonproliferation: Assessing Nuclear Power and Its Alternatives

Nonproliferation Policy Education Center

Nonproliferation Policy Education Center
Arlington, VA 22216
www.npolicy.org

Printed in the United States of America.

ISBN: 978-1-7371113-4-4

Cover image: Solar panels before a nuclear power plant and wind turbines (credit: Adobe Stock).

Nonproliferation Policy Education Center

The Nonproliferation Policy Education Center (NPEC), a 501(c)3 nonprofit organization, is a nonpartisan, educational organization founded in 1994 to promote a better understanding of strategic weapons proliferation issues. NPEC educates policymakers, journalists, and university professors about proliferation threats and possible new policies and measures to meet them.

For current publications of the Nonproliferation Policy Education Center,
please visit www.npolicy.org.

Contents

Acknowledgments

This research was supported by generous grants from the Sarah Scaife Foundation and the Smith Richardson Foundation.

Net-Zero and Nonproliferation: Assessing Nuclear Power and Its Alternatives

Henry Sokolski

Six years ago, NPEC ran a mock execution of a law Congress passed in 1978 but that the Executive refused to implement — Title V of the Nuclear Nonproliferation Act of 1978. Title V called on the State and Energy Departments to conduct country-specific analyses of how developing states might best meet their energy needs without nuclear power. It also called for the creation of an energy Peace Corps and an assessment of what our government was spending on energy development aid-related projects.

The gist of this law was best captured by the language contained in Title V's opening section:

> The United States shall endeavor to cooperate with other nations, international institutions, and private organizations in establishing programs to assist in the development of non-nuclear energy resources, to cooperate with both developing and industrialized nations in protecting the international environment from contamination arising from both nuclear and non-nuclear energy activities, and shall seek to cooperate with and aid developing countries in meeting their energy needs through the development of such resources and the application of non-nuclear technologies consistent with the economic factors, the material resources of those

> countries, and environmental protection. The United States shall additionally seek to encourage other industrialized nations and groups of nations to make commitments for similar cooperation and aid to developing countries. The President shall report annually to Congress on the level of other nations' and groups of nations' commitments under such program and the relation of any such commitments to United States efforts under this Title. In cooperating with and providing such assistance to developing countries, the United States shall give priority to parties to the Treaty.[1]

Title V aimed to restrain and counterbalance the global promotion of nuclear power. It recognized the inherent nuclear weapons proliferation danger that any expansion of nuclear power posed. As Leonard Weiss, a key author of Title V noted, "a nonproliferation policy worthy of the name should begin with recognition of the inherent weapons proliferation potential in the technology and avoiding encouragement of its adoption whenever economically viable alternatives are available."[2]

When NPEC started its efforts to write up this and the other required Title V reports, the staff on the Senate Foreign Relations Committee asked to see what NPEC produced to use it to pressure the Executive finally to implement the law. NPEC commissioned a number of studies on how Iran, Saudi Arabia, China, and Taiwan might best meet their energy requirements without nuclear power. The center also contracted studies on the history and intent of Title V and on what government programs were already in play that aligned with Title V's stated objectives. As soon as NPEC's project was completed, the Senate Foreign Relations Committee staff prepared

---

1. See Appendix of this volume, "Text of Title V of the Nuclear Nonproliferation Act of 1978," Section 501, p. 364.
2. See Leonard Weiss, "A Brief History of Title V in the Nuclear Nonproliferation Act of 1978," pp. 11.

a letter to Secretary of State Hillary Clinton asking the Secretary finally to implement Title V and file the reports required by law.

Then, something unexpected occurred. The committee's legal counsel discovered that the Secretary was under no obligation to comply: Congress had eliminated Title V's reporting requirements along with several hundred other Congressionally mandated reports back in 1995. This was embarrassing. Flummoxed, I quietly set the book manuscript aside.

Why, then, release it today? Despite the project having been completed nearly six years ago, its country-specific energy assessments — on Taiwan, Saudi Arabia, China, and Iran — are now more timely than when they were written. In October, the Biden Administration announced it is still considering extending civilian nuclear cooperation with Riyadh[3] that would allow the Kingdom to enrich uranium — a process that can bring states within weeks of acquiring the bomb. Administration officials no longer question if Saudi Arabia really needs nuclear energy to meet its energy requirements. Shouldn't they?

Meanwhile, Taiwan's presidential election this January will, among other things, decide if Taiwan will build more nuclear reactors or not. Again, how necessary building new plants is a central question. As for China, the Pentagon has become increasingly concerned that the two "peaceful" fast breeder reactors and plutonium reprocessing plants Beijing is building will be used to make hundreds of bombs worth of weapons plutonium. One of the two fast breeder reactors is already operating. The question these dangerous nuclear activities raise is just how necessary are they to meet China's energy requirements.

---

3. See Brit Farmer, "Biden: Saudi-Israeli Normalization Not Dead" *60 Minutes*, October 15, 2023, available at https://www.cbsnews.com/news/biden-saudi-arabia-israel-normalization-not-dead-60-minutes/.

Then, there's Iran, which is intent[4] on building reactors of Iranian design. It plans on expanding its nuclear power program from roughly one gigawatt electrical capacity to 11. Given Iran's renewables potential and oil and gas reserves, how much sense does such nuclear construction make?

The studies in this volume answer these questions and more. In addition to country-specific energy studies for China, Taiwan, Iran, and Saudi Arabia, it includes a brief legislative history of Title V along with a rough inventory of energy assistance programs the U.S. government conducts relevant to Title V's objectives.

It was timely 45 years ago. It was timely six years ago. It is still timely today. Just this last month, the United States pledged to achieve net-zero global carbon emissions by 2050.[5] To achieve that requires country-specific assessments of how best to meet this goal in a manner almost identical to that stipulated in Title V. The country-specific studies contained in this volume offer up worthy models.

Meanwhile, the Energy Department just created a Clean Energy Corps to accelerate efforts to clean up America's energy sector.[6] This corps' goals are similar to those of the energy Peace Corps that Title V proposed. The key difference is that the Clean Energy Corps is

---

4. "Iran begins procedures to build indigenous nuclear reactor," Nuclear Energy International, 11 May 2022 available at https://www.neimagazine.com/news/newsiran-begins-procedures-to-build-indigenous-nuclear-reactor-9687300.

5. See Dipanita Nath, "Global Renewables and Energy Efficiency Pledge at COP28: what is the tripling renewable energy target?" India Express, December 10, 2023, available at https://indianexpress.com/article/explained/explained-climate/cop28-tripling-renewable-energy-target-9062282/.

6. See "Biden launches US 'climate corps' amid pressure from environmental advocates," Al Jazeera, September 20, 2023 available at https://www.aljazeera.com/news/2023/9/20/biden-launches-us-climate-corps-amid-pressure-from-environment-advocates and US Department of Energy, "Clean Energy Corps," available at https://www.energy.gov/CleanEnergyCorps.

currently focused only on domestic American energy endeavors. Yet, the United States emits 14 percent of the world's carbon dioxide.[7] If the Energy Department is serious about achieving net-zero emissions, expanding the Clean Energy Corps' authority to cover U.S. energy assistance to friendly developing states would necessarily make sense as would a Title V-like report on the feasibility and advantages of creating a U.S. energy Peace Corps like entity what would work both domestic and overseas clean energy development.

Finally, Washington officials also have committed to tripling global nuclear power and renewables[8] three-fold to help achieve net-zero carbon emissions by mid-century. This immediately raises a practical question: If the United States is serious about its net-zero pledge, what is the quickest, cheapest way to achieve it? Would it be quicker and cheaper to invest in natural gas conversions of coal plants as a bridge to moving entirely carbon-free forms of energy or to skip this gas transition and immediately invest all of our resources into renewables? What of nuclear power, which currently is quite capital intensive? Does it make more sense to build many of these plants now when the cost of borrowing is so high or should such investments be delayed until the overnight construction and borrowing costs decline?

To answer these issues, in turn, requires clarity on two fronts. First, what are the real overnight and levelized costs for renewables, nuclear, fossil-fueled energy systems, and their related energy storage and distribution systems? Unfortunately, solid information

---

7. *Statista,* "Emissions in the U.S.," December 23, 2023, available at https://www.statista.com/topics/3185/us-greenhouse-gas-emissions/#topicOverview.
8. U.S. Department of Energy, "At COP28, Countries Launch Declaration to Triple Nuclear Energy Capacity by 2050," December 1, 2023, available at https://www.energy.gov/articles/cop28-countries-launch-declaration-triple-nuclear-energy-capacity-2050-recognizing-key.

of this sort is all too scarce. Rarer still are sound, official government figures on these costs. The reason why is simple: government officials and industry have little incentive to clarify the ways different kinds of energy systems are subsidized and how large these subsidies are. Instead, they support an "all-of-the-above" approach to handing out government energy support.

This is unnecessarily wasteful. If Congress or the Executive wants to cut this waste, though, routine, up-to-date, official reports comparing the levelized and overnight construction costs of different energy systems and their related distribution and storage systems are essential. These costs, of course, must factor in government subsidies.

Second, one or more greenhouse gas cost abatement curve models need to be made available to help determine how much and when investments in different energy types and their related storage and distribution systems would make the most sense to get to net-zero quickest at the least cost. Currently, the best known model is the one the McKinsey Company has developed.[9] Both environmentalists and nuclear utilities have used this model. Other abatement curve models are available.[10] It would be useful for the U.S. government to endorse one or more of these, make them publicly available, and use them to devise investment strategies to achieve net-zero in the cheapest, quickest fashion.

Finally, just as Title V requires under Section 501, the Departments of Energy and State should catalogue what the U.S. government is

---

9. See McKinsey & Company, "Pathways to a Low-carbon Economy: Version 2 of the Global Greenhouse Cost Abatement Curve, 2009, available at https://www.energy.gov/articles/cop28-countries-launch-declaration-triple-nuclear-energy-capacity-2050-recognizing-key.
10. For a brief overview and discussion of several cost abatement curve models see Wikipedia, "Marginal Abatement Cost," available at https://en.wikipedia.org/wiki/Marginal_abatement_cost. Note well that many other models have been devised besides those listed in this article.

doing with other nations, international institutions, and private organizations to meet their energy requirements in the cleanest fashion possible. This volume includes a rough catalogue of such programs. It, however, needs to be expanded and updated.

A Brief History of Title V in the Nuclear Nonproliferation Act of 1978

Leonard Weiss

On March 3, 1977, Senator Charles Percy (R-Ill) introduced a bill in the U.S. Senate entitled the Nuclear Nonproliferation Act. The main co-sponsor of the bill was Senator John Glenn (D-OH), whose nuclear subcommittee staff worked cooperatively with Percy's staff to produce the bill. The bill was referred for consideration to three Senate Committees; Governmental Affairs, Foreign Relations, and (for part of it) the Energy Committee. At the Governmental Affairs Committee markup of the bill the following September, Glenn introduced a major rewrite and reorganization of the bill as an amendment which was unanimously adopted at the markup. Glenn's amendment was the result of three months of intense negotiation between myself as Glenn's nuclear subcommittee Staff Director and representatives of four government departments and agencies representing the interests and policies of the Carter Administration (State, Energy, ACDA, and the NRC).

The Carter Administration had produced its own nonproliferation bill in April 1977, and its content was promoted by administration negotiators at the markups of the Percy Bill in the three committees to which Percy's original bill had been assigned. While the administration bill, like the Percy bill, reflected the desire for more controls over the production of separated plutonium from US-supplied nuclear fuel or from spent fuel in US-exported reactors, it did not question the wisdom of expanding nuclear power technology to places where it was not already entrenched.

New export controls presented a sensitive issue for the international nuclear industry and the governmental bureaucracies that had grown up around it and supported it. A direct challenge to the expansion of nuclear power was politically sensitive both domestically and internationally, and at the time (the mid-70s) would have gone against the prevailing consensus about the future role of nuclear energy in the development of the world's energy resources. Nonetheless, I believed a nonproliferation policy worthy of the name should begin with recognition of the inherent weapons proliferation potential in the technology and avoiding encouragement of its adoption whenever economically viable alternatives are available.

National security considerations in any country perceiving an external threat would raise the temptation in that country to turn nuclear power technology, if available, toward weapons. Indeed, any defense minister doing his or her job would be derelict in duty if he or she did not examine all possibilities, including nuclear ones, for addressing a perceived national security threat. Having nuclear technology available via a national power program would be a natural candidate for weapon development, the NPT notwithstanding. Non-nuclear alternatives in energy assistance provided by advanced countries like the United States would thus be preferable from a nonproliferation perspective. This was made manifest in the Glenn rewrite of the Percy bill as follows:

First, a statement in the original Percy bill declaring it U.S. policy to aid other nations in identification and adaptation of "appropriate" energy production technology was strengthened and made more explicit in the Glenn rewrite by explicitly clarifying appropriate technologies to mean solar and other unconventional technologies and, in particular, to identify alternative options to nuclear power in aiding nations to meet their energy needs, consistent with the economic and material resources of those nations and environmental protection.

This statement of policy was made legislatively explicit by adding a new section (Title V) to the bill in the markup requiring the United States to "endeavor to cooperate with other nations, international institutions, and private organizations in establishing programs to assist in the development of non-nuclear energy resources, to cooperate with both developing and industrialized nations in protecting the international environment from contamination arising from both nuclear and non-nuclear energy activities, and shall seek to cooperate with and aid developing countries in meeting their energy needs through the development of such resources and the application of non-nuclear technologies consistent with the economic factors, the material resources of those countries, and environmental protection."[11] The meaning and intent of this language was reinforced by the description of Title V in the report issued on behalf of the two senate committees to which the legislation was referred for consideration:[12]

"DOE – under the general policy guidance of State and in cooperation with the Agency for International Development (AID) – is directed to initiate a program for the exchange of U.S. scientists and energy experts with those of developing countries."

---

11. Nuclear Nonproliferation Act of 1978, H.R. 8638, 95th Cong. (1978)
12. See Senate Report No. 95-467, 95th Congress *Ist Session,* to accompany S.897--- reported by the Committee on Government Affairs, the Committee on Foreign Relations, October 3, 1977 [also available as Committee Print: Legislative History of the Nuclear Nonproliferation Act of 1978 H.R.8638 (Public Law 95-242) prepared for the Subcommittee on Energy, Nuclear Proliferation, and Federal Services of the Committee on Governmental Affairs, United States Senate, by the Congressional Research Service Library of Congress, January 1979; pp 482-483].

Funds were authorized for carrying out this program. The State Department was directed to coordinate activities under this Title with those carried out under the Foreign Assistance Act. Another section of Title V required the President to report within one year on the feasibility of expanding these cooperative activities into an international effort to include a "scientific peace corps" designed to encourage large numbers of technically trained volunteers to live and work in developing countries. As the report stated:

> "Title V stems from the recognition that the first step in any nonproliferation strategy aimed at developing countries should be to cooperate with and aid such countries in identifying non-nuclear alternatives in meeting their energy needs. In general, countries that meet their energy requirements through indigenous, non-nuclear resources should be encouraged to do so consistent with environmental considerations. Past U.S. policies have emphasized nuclear energy- a natural consequence of our commitment to spread the benefits of nuclear power throughout the world and to fulfill Article IV of the NPT which calls for the fullest possible exchange of nuclear technology with due consideration for the needs of developing countries. The spread of nuclear technology was also facilitated by low-interest Eximbank loans for the purchase of reactors as well as extensive technical training in nuclear technology for foreign students."
>
> "Title V is designed to balance these policies by offering cooperation and assistance in developing indigenous non-nuclear energy technologies with priority being given to NPT Parties. In addition to the contribution which these efforts would make to nonproliferation, such assistance will promote political and economic stabilization in developing countries through reduction of their dependence on foreign oil and highly capital-intensive technology and will accelerate the availability and

> utilization of renewable energy technology … with accompanying technological improvements."

In carrying out the cooperative program described in Title V, it was felt that programs underway by AID and DOE that are consonant with the aims of Title V were inadequate in terms of scope and management and that long-term success would be enhanced by encouraging explicit scientific cooperation and exchanges between energy specialists in the U.S. and those in developing countries through the "scientific peace corps" idea.

As indicated, the intent of Title V was to help move energy assistance to developing countries away from nuclear power. The antipathy of the Energy Committee to this idea in their markup of the bill reflected the attitude of the nuclear industry, which interpreted Title V, as well as other sections of the bill, as a frontal assault on nuclear expansion.

To help construct a counter-argument to this interpretation, I incorporated into the bill language intended to address a problem vexing the industry at the time, the inability to guarantee the delivery of nuclear fuel for the lifetime of exported reactors. Without such guarantees, the reactor export market would be operating under a cloud of uncertainty unavoidable for a technology requiring 10-15 years for construction of a reactor and 40 years of operational lifetime. While it was understood that actually providing iron-clad guarantees of fuel delivery was impossible and would not necessarily be desirable in a bill focused on nonproliferation, I believed that a symbolic expression of support along that line might mollify the industry's supporters sufficiently to at least not actively oppose the bill.

Accordingly, I wrote into the bill a provision urging the administration to ensure ongoing appropriate nuclear enrichment capacity by the U.S. and to create an international uranium fuel contracting and delivery system via establishment of an international fuel authority or INFA (an idea contained in an earlier freestanding Glenn bill (S. Con. Res. 17, introduced on April 1, 1977)), with international management that might be less subject to the domestic political winds of the countries supporting INFA. I put this provision right up front in the reorganized nonproliferation bill as Title I, thereby underscoring the need for Title V as a balancing provision offsetting the otherwise pro-nuclear bias in U.S. policy. This was important since the Energy Committee was hostile to Title V and recommended striking it in the Energy Committee markup of the bill. However, there was no attempt to strike the provision when the bill came to the Senate floor for action. Title I may have played a helpful role in that case. Unfortunately, although the Carter Administration supported the provision, no effort was made to implement it in accordance with its philosophy, and that has been the case with all succeeding administrations.

Although it was not explicitly expressed in the legislation, implementation of Title V philosophy should mean restraint, if not opposition, toward support for nuclear power in more advanced countries as well, since proliferation issues arise at every level of nuclear technological development. The current belief in some influential quarters that climate change requires more extensive use of nuclear power means that there is likely to be more proliferation among developing countries in the future.

By focusing on solar and other renewable energy sources in assisting developing countries to meet their energy needs, Title V makes a contribution not only to nonproliferation, but also to the alleviation of climate change and to the protection of the environment from nuclear waste. This was recognized by at least one of the founders of the International Renewable Energy Association (IRENA), whose inspiration, I was told, to create the association was triggered by Title V.

U.S. International Energy Assistance and Cooperation: Has the Title V Objective Been Realized?

Robert Ichord, Jr.

I. Executive Summary

Energy has been recognized by both the Obama and Trump Administrations as an important element of U.S. foreign policy and national security strategy. But the clean energy/climate change approach under Obama has given way to the fossil fuel priority and "energy dominance" theme of the Trump Administration's National Security Strategy.

Title V of the Nuclear Nonproliferation Act of 1978 authorized a program to help developing countries pursue non-nuclear energy alternatives. Although it was never specifically funded, the U.S. government has pursued an active international energy assistance and cooperation program with developing countries for many years.

The global energy role of developing countries has greatly expanded since 1978, with non-OECD countries now accounting for almost 60% of global energy consumption. With their income, population, and urban growth, they are projected to account for all of the world's growth in primary energy from 2017-2040.[13] These are the big energy markets of the future and the fundamental question is: how

13. See both International Energy Agency, World Energy Outlook 2017, p. 65 and BP Energy Outlook 2018, p. 14.

will the U.S. government and its companies pursue these opportunities and face the challenges of China, Russia, Japan, Germany, S. Korea and other competitors?

This paper presents a snapshot of current USG international energy programs and the roles of the main USG agencies in implementation during the Obama Administration and the first year of the Trump Administration. It distinguishes four major types of activity: (1) technical assistance for energy policy and sector reform; (2) energy technology cooperation; (3) energy capital project assistance; and (4) energy trade and investment promotion.

U.S. agency funding in FY16 (not including the Export-Import Bank or Treasury funding to the International Finance Institutions) is estimated at over $2.5 billion, with the largest outlays from the Overseas Private Investment Corporation, the Millennium Challenge Corporation, and the U.S. Agency for International Development. The main geographical focus of these programs and projects was in Sub-Saharan Africa and the Near East (including Afghanistan and Pakistan). One of the main Obama Initiatives was Power Africa, a "whole of government" initiative coordinated by the NSC and USAID that focused on mobilizing U.S. private investment for power generation and improving energy access in Sub-Saharan Africa. This initiative gained substantial bipartisan support in Congress, which led to the passage of the Electrify Africa Act of 2015.

The incoming Trump Administration eliminated the emphasis on climate change and proposed deep cuts in FY18 and FY19 foreign operations budgets. Congress restored much of this funding in FY18 and FY19, although the overall $54 billion appropriated in both years was below the peak FY17 level of $57 billion.[14] Preliminary

---

14. Congressional Research Service, Department of State, Foreign Operations and Related Programs," FY19 Budget and Appropriations, p. 2.

energy program numbers for FY17 appeared to show a reduction in funding to around $2 billion, reflecting drops in USAID, MCC, and OPIC obligations. Although President Trump originally proposed the termination of the Overseas Private Investment Corporation (OPIC) and the Trade and Development Agency (TDA), the Administration changed its position in the Fall of 2017 and supported the creation of a new Development Finance Corporation which would expand OPIC authorities and incorporate USAID's loan guarantee program (DCA). The President signed the Build Act (PL 115-254) on October 5, 2018, creating the new International Development Finance Corporation (USDFC) with a mandate to promote private investment in energy and other development sectors and carry out feasibility studies as well as expanded financing and insurance activities.[15]

Another Trump Administration Initiative has been the "Free and Open Indo-Pacific Strategy," which includes a component for energy (Asia EDGE – Enhancing Development and Growth through Energy). At the time of this writing, the details and funding parameters for this program have not been announced.

While these new initiatives appear to be potentially important for future U.S. international energy cooperation, this paper argues for the development of a new high-priority, global programmatic emphasis and framework that would restore an "all of the above" energy emphasis to the foreign operations budget and bring a coherent approach among the National Security Council, the new U.S. Development Finance Corporation, U.S. Agency for International Development, Trade and Development Agency, Millennium Challenge Corporation, and U.S. Export-Import Bank as well as the Departments of State, Energy, Treasury and Commerce. The Power Africa experience represents a model that

15. See https://www.opic.gov/sites/default/files/files/BILLS-115hr302_BUILDAct2018.pdf for a text of the Build Act.

could be adapted to other regions to ensure support for U.S. security, economic/commercial and environmental interests.

II. Background and Purpose

Title V of the Nuclear Nonproliferation Act of 1978 authorized but did not fund the creation of a program to help developing countries pursue non-nuclear energy alternatives.[16] The United States Government has, however, pursued international energy assistance and cooperation programs with developing countries over the years since 1978 that have focused on non-nuclear energy development. These programs have been driven by security, economic development, and environmental objectives but have generally, except in a few countries of the former Soviet Union, not been motivated by nuclear nonproliferation concerns or desires to limit nuclear power development. Nonproliferation remains nevertheless an important strategic foreign policy concern, especially as additional developing countries seek to develop nuclear capabilities and power facilities and China and Russia step up efforts to finance and build nuclear plants in the politically-sensitive countries in the Middle East and South Asia and challenge U.S. leadership in the international nuclear order.

The global energy scene has changed dramatically since 1978 with the emergence of the developing, non-OECD world as the leading region in terms of energy consumption and growth and the impact of changing economic and environmental factors on energy technology choices. The non-OECD countries in 2016 accounted for

---

16. Title V specifically states: "The United States shall initiate a program, consistent with the sims to cooperate with developing countries for the purpose of – (1) Meeting the energy needs required for the development to such countries; (2) Reducing the dependence of such countries on petroleum fuels, with emphasis given to utilizing solar and other renewable energy resources; (3) Expanding the energy alternatives available to such countries."

58 percent of global primary energy consumption, compared to only 42 percent in 1980.[17] Consumption has grown by almost twice and CO2 emissions have risen from 15GT to over 32GT.[18] Although fossil fuels still account for around 80% of global primary energy consumption, the costs of renewables have dropped rapidly, and renewable energy accounted for more than half of all the new electricity generation capacity additions in 2016.[19]

The purpose of this report is to consider current U.S. international energy assistance and cooperation efforts with developing countries. It will principally review the nature, magnitude and direction of policies, strategies and programs under the two Obama Administrations (2009-2016), and the first year of the Trump Administration (2017). The report does not propose to assess and evaluate the effectiveness and impact of the energy assistance and cooperation activities and is mainly descriptive in nature. The actions and strategies of the Trump Administration clearly show a substantial change in direction from the Obama Administration's strong emphasis on global climate change and renewable energy. The report will speculate on how these changes may affect future international energy programs and make some suggestions on what Congress might consider in authorizing new activities in this sphere.

III. Goals, Strategies, and Mechanisms

The important role of energy in our foreign policy has been evident in both the Obama and Trump Administrations' policies. But the U.S. energy position has dramatically changed since 2010, when Obama's National Security Strategy stressed U.S. energy

17. International Energy Agency, World Energy Outlook 11, Table 2.3, p. 81 and BP, Statistical Review of World Energy June 2017, p. 8.
18. IEA, World Energy Outlook 2017, Figure 1.1, p. 42.
19. Ibid., p. 230.

dependency and the need to take the lead in the development of clean energy technologies that can cut greenhouse gas emissions, improve energy efficiency, and reduce dependence on oil.[20] This focus was highlighted in the first Quadrennial Diplomacy and Development Review (QDDR) in 2010, which emphasized that "international energy policy lies at the intersection of economics, geopolitics and development,"[21] and proposed the creation in the Department of State of both an Undersecretary for Economic Growth, Energy and Environment and a new Bureau of Energy Resources "to unite our diplomatic and programmatic efforts in oil, natural gas, coal, electricity, renewable energy, energy governance, strategic resources and energy poverty."[22]

With the shale oil and gas revolution in the United States fundamentally changing the U.S. oil and gas import position, the Trump Administration pledged to "embrace energy dominance" in its December 2017 National Security Strategy.[23] This strategy calls for five main energy-related actions:

- Reduce barriers to U.S. energy development;
- Promote exports of energy, technology, and services;
- Ensure energy security of the U.S. and its allies;
- Attain universal energy access;
- Further America's technological edge in energy.

The Trump strategy as it has emerged is very fossil-energy dominated and the Administration has touted U.S. LNG, coal, and

20. The White House, US National Security Strategy, 2010, p. 30.
21. US Department of State, Quadrennial Diplomacy and Development Review, 2010, p. 40.
22. Ibid., p. vii.
23. The White House, US National Security Strategy, December 2017, pp. 21-2.

oil exports and their benefits to both the U.S. and our allies. It stands in sharp contrast to Obama's 2015 National Security Strategy that cited climate change as a top risk to U.S. security. A January 11, 2018 letter by 160 members of Congress to President Trump calls attention to this omission and the importance of climate change to U.S. national security.[24]

A signature initiative of President Obama was the Global Climate Change Initiative (GCCI), which was included as part of the September 22, 2010, Policy on Global Development. GCCI sought to integrate climate change into U.S. foreign assistance strategy to foster a "low-carbon future."[25] This initiative was founded on the imperative of international cooperation such as the U.S. had fostered in creating the Major Economies Forum for Energy and Climate on March 28, 2009. The U.S. joined with other developed countries following the Copenhagen Conference of Parties in committing resources to climate change and clean energy. The U.S. "Fast Start Climate Finance Initiative" ran from 2010-2012 and involved U.S. assistance to more than 120 countries for clean energy as well as adaptation and sustainable landscapes. These resources, which included bilateral assistance, trade and investment, and financing to multilateral and International Financial Institutions totaled an estimated $2.435 billion over the 2010-2012 period.[26] The U.S. also worked to establish the Green Climate Fund and pledged $3 billion, the most of any country, to this new mechanism. Overall the industrial countries endorsed the goal of mobilizing $100 billion per year for developing countries' climate projects by 2020.

---

24. Phil McKenna, "106 Lawmakers Urge Trump: Restore Climate Change in National Security Strategy," Inside Climate News, January 16, 2018.
25. See Obama Archives, Climate and Energy, available from www.obamawhitehouse.archives.gov.
26. US Department of State, "Meeting the Fast Start Commitment, FY12, available from *2009-2017.state.gov/documents/organization/177990.pdf*.

Underlying this international effort was the U.S. commitment to reduce U.S. greenhouse gas emissions. The first pledge in November 2009 called for a 17% reduction from 2005 levels by 2020. Subsequently, prior to the Paris Climate Summit, the U.S. increased the pledge to 26-28% reductions from 2005 by 2025. The Administration introduced the Clean Power Plan in August 2015, which further proposed a 32% reduction from 2005 levels by 2030.

With the election of Donald Trump, U.S. climate policy dramatically changed. Trump announced U.S. plans to withdraw from the Paris Agreement on June 1, 2017; and reneged on the remaining $2 billion of the $3 billion Obama pledge to the Green Climate Fund. In October 2017, EPA Administrator Pruitt announced the decision to terminate the Clean Power Plan.

During the Obama Administration, State and DOE actively supported and participated in the UN's Sustainable Energy for All Initiative, championed by UN Secretary General Ban Ki Moon and co-chaired with the President of the World Bank Jim Yong Kim. This initiative involved a partnership among governments, IFIs, and private sector organizations focused on three main goals: universal access to electricity and clean cooking fuels by 2030; doubling of the rate of energy efficiency by 2030; and doubling of the role of renewable energy in the global energy mix by 2030. Secretary Kerry was a member of the High Level Advisory Group and State served on the Steering Committee. The work of this initiative provided the foundation for the Sustainable Development Goal Seven on energy which was endorsed by world leaders along with other SDG goals at the UN in the fall of 2015. With the departure of Ban Ki Moon, the SE for All Initiative was spun off to a non-governmental organization based in Austria, where it receives financial support from the Austrian government and European donors. The Trump NSS does mention promoting universal energy access but the approach has not been clearly articulated. Power Africa with its focus on increasing energy access to 30 million consumers and "beyond the grid" opportunities, has however survived administrations and is cooperating with SE For All.

The Obama strategy clearly emphasized the importance of USG interagency cooperation and working through international alliances and organizations, particularly on global issues like climate change. The PEPFAR health experience was commonly cited as an example of the "Whole of Government Approach" that they espoused. President Obama's Power Africa Initiative, which will be discussed below, embodied this Whole of Government Approach and has been a major instrument of U.S. international energy assistance. It has brought together a wide range of U.S. government departments and agencies within a common coordination framework: The National Security Council and Office of Management and Budget, Department of State, U.S. Agency for International Development (which provided day-to-day management from an office in South Africa), Millennium Challenge Corporation, Departments of Energy, Treasury and Commerce, Trade and Development Agency, Overseas Private Investment Corporation, Export-Import Bank, and the Environmental Protection Agency. The sections below describe the roles and activities of these various agencies in the U.S. international energy approach, focusing on the energy technical assistance, energy technology cooperation, energy capital assistance, and energy trade and investment promotion dimensions. Figure 1 shows key U.S. agencies and their functions.

Figure 1: Role of U.S. Agencies in International Energy Cooperation

US AGENCY INTERNATIONAL ENERGY PROGRAMS

Agency	Type of Program	Focus of Activities
Department of State	--energy policy and investment dialogue	--oil and gas diplomacy -- power sector investment
Agency for International Development	--technical assistance for energy sector reform --selected capital assistance	--energy regulation --power system improvement --renewables and off-grid
Department of Energy	--technology cooperation --energy assessments, policy dialogue	--energy efficiency, LNG, CCS, renewables, environmental management
Millennium Challenge Corporation	--grant capital assistance --related technical assistance	--power distribution/transmissions --small hydro and renewables
Trade and Development Agency	--feasibility studies, trade missions and promotion events	--gas strategies, power and renewables
Overseas Private Investment Corporation	--loans, equity, insurance	--power generation, renewables
Export-Import Bank	--loan guarantees, export credits	--oil and gas equipment and services, --power sector equipment and services

IV. Technical Assistance for Energy Policy and Sector Reform

The United States provides a broad range of energy technical advisory services and training to developing countries to help them build the institutional and human resource capacities to plan and manage better their energy sectors.

The U.S. Agency for International Development (USAID) is the principal vehicle for U.S. energy technical assistance although MCC, State, and DOE are also important providers.

The developing countries face a range of challenges in the management and transition of their energy sectors, ranging from basic institutional, governance, and human resources constraints to critical energy access, poverty, and environmental issues.[27] USAID seeks to address these problems through global, regional and country programs and projects. Creating the right policy and institutional framework (the "enabling environment") for private investment and entrepreneurial development is at the core of USAID programs. Figure 2 below provides a snapshot in FY16 of the main project vehicles that USAID has developed in the sector at the various levels and in major areas of programmatic emphasis. Projects are grouped here in four major areas, although projects often cut across the areas:

(1) Institutional and Regulatory Development: This area includes broader sector reform projects and ones that are focused on development of professional energy regulatory capacity and an investment framework conducive to private investment and sound utility operation.

(2) Energy Efficiency and Renewable Energy Development: Projects in this area generally have been called "Clean Energy" projects and seek to promote low emission options, assess resources and develop strategies and support national programs to advance energy efficiency and renewable energy technologies and their commercialization.

(3) Energy Access and Rural Poverty: The challenge of expanding energy access, one of the key elements of the UN Sustainable Development Goal 7, is a central USAID concern and is closely

---

27. For a discussion of these challenges, see Robert F. Ichord, Jr., "Transforming the Power Sector in Developing Countries: A Strategic Framework for Post-Paris Action," Atlantic Council, Global Energy Center, October 2016, available from *https://www.atlanticcouncil.org/images/publications/Transforming_the_Power_Sector_in_Developing_Countries_web_1020.pdf.*

linked with poverty reduction and how energy can increase rural productivity and incomes.

(4) Utility Performance Improvement and Grid Modernization: A major program focus has been on enhancing the performance, both financial and technical, of the national, and in some countries municipal, utility systems. Modernization of electricity grids and the issues of facilitating integration of variable renewable energy sources is an important component of this work.

Figure 2: Illustrative Description of USAID Energy Programs

USAID ENERGY PROGAM DIRECTIONS AND KEY PROJECTS

	Institutional & Regulatory/Investment	Energy Efficiency & Renewable Energy Dev.	Energy Access & Poverty Reduction	Utility Performance Improvement
Global Projects	--Regulatory Partnership --CTI Private Financing Advisory Network	-- Energy Efficiency for Clean Development -- EC-LEDS (Low Emissions)	--Powering Agriculture	--Utility Partnership --Sector Reform and Utility Commercialization
Regional/Multi-Country Projects	--Power Africa --Caribbean Clean Energy Finance Facility --South Asia Regional Initiative	--Clean Power Asia	--Africa: Scaling Off-Grid Energy --Developing Sustainable Cookstove Sector --Renewable Micro-finance	--Engendering Utilities
Country Projects	--Congo(DRC) Power Sector Reform --Jordan Energy Sector Capacity Building --Kosovo Energy Reform --Philippines Energy Policy	--Colombia Clean Energy --India– Partnership to Accelerate Clean Energy --Indonesia Clean Energy --Vietnam Clean Energy	--Bangladesh: Catalyzing Clean Energy --Haiti: Pilot Project for Sustainable Electricity Distribution	–Colombia Clean Energy -- Nepal Hydropower --Pakistan Hydro, Thermal Rehabilitation Projects and Wind Transmission --Ukraine: Municipal Energy --Afghanistan: Kajaki Dam

An exact accounting of U.S. energy technical assistance is difficult to provide since the individual projects may fall under different energy, economic, infrastructure and environmental budget line items. This report focuses on obligations of funding in actual contracts or grants. These actual outlays are usually different than budget requests to Congress or country budget allocations for different sectors. Contracts and grants for a specific project are often funded in tranches over several years. Annual obligations numbers, therefore, may only capture a portion of the funding planned or ultimately provided to a project. Based on USAID's relatively new foreign assistance dashboard,[28] USAID obligated $446 million for energy in FY16 and was active in 26 countries. Removing the capital assistance projects in Afghanistan and Pakistan and adding clean energy and low emissions projects listed under the environment category, a rough estimate of USAID's energy technical assistance obligations is in the range of $300 million in FY16. The largest country specific energy obligations (excluding global and regional projects) were in: Afghanistan, Pakistan, Liberia, Ukraine, Jordan, Congo DRC, Bangladesh, India, Colombia, and the Philippines. Figure 3 shows estimated obligations by region, with the Near East (including Afghanistan and Pakistan) and Africa accounting for over 75% of the total energy obligations.

28. See Foreign Aid Explorer Dashboard, available from https://explorer.usaid.gov.

Figure 3: Estimated USAID Energy Program Funding in FY16

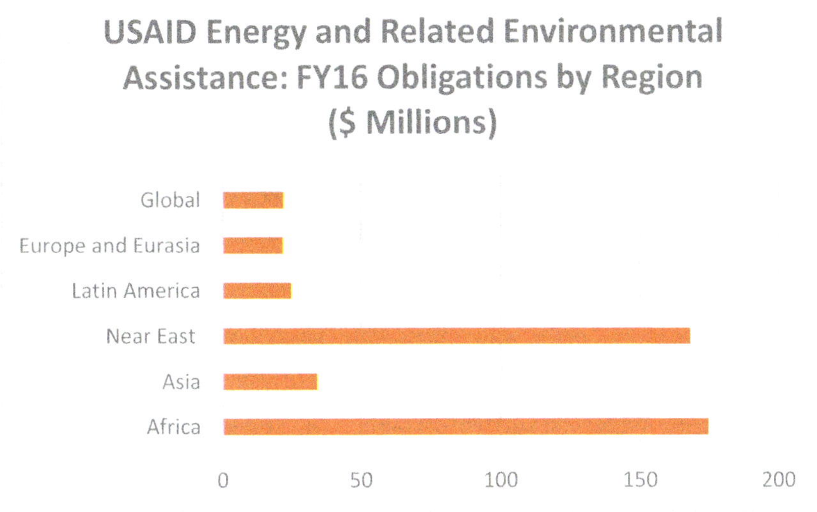

Source: USAID Foreign Assistance Dashboard for FY16 (https://explorer.usaid.gov.): See Appendix B for Details

Under the Obama Administration, some of the energy assistance funding was grouped under the Clean Energy pillar of the GCCI initiative and a centralized decision process involving State's climate offices had a major role in determining country levels and the role of clean energy vis-à-vis adaptation and sustainable landscapes funding. Most of this clean energy funding came from the Development Assistance (DA) Account. Considerable energy funding however was programmed outside of the GCCI, including Economic Support Fund activities in Afghanistan, Pakistan, and Jordan and AEECA funds for programs in the Europe and Eurasia region.

A major initiative under the GCCI was the EC-LEDs (Enhancing Capacity in Low Emissions Development), which worked with 25 partner countries on the development and implementation of low emissions development strategies and programs. The program was led by USAID with technical coordination by DOE's National Renewable Energy Laboratory (NREL). These efforts sought to be inter-Ministerial and cross sectoral in their engagement with counterparts and applied a planning methodology that examined possible low emission policy options in relationship to a business as usual scenario. Of the 25 EC-LEDs countries, about 11 had a major concentration on clean energy. In many cases, the main counterpart was the Ministry of Environment, which in most countries was relatively weak politically compared with the Ministry of Energy.

Two other energy partnership programs that fostered regulatory and electric utility system reforms have been pursued under Cooperative Agreements with the National Association of Regulatory Utility Commissioners (NARUC) and the U.S. Energy Association (USEA). These programs involved professional exchanges and seminars/workshops that brought together U.S. and country government and energy industry counterparts. These programs generally complemented bigger USAID or State technical assistance contracts with leading U.S. consulting-company consortia that provided both long and short-term experts to work in depth on policy, regulatory, and institutional reform areas.

A major Presidential initiative under the Obama Administration that, except for a small amount of GCCI funds, was funded separately as Power Africa. This ongoing program works with African countries to address their serious deficit of energy by expanding investment for generation and increasing access to electricity for businesses and households. It both fosters energy sector reform as well as promotes private sector investment. It targets in on removing specific obstacles that can facilitate and unlock specific transactions. It followed the "Whole of Government" model, was coordinated by the NSC as well as USAID in the field from an office in South Africa and included significant involvement from U.S. trade and investment agencies, i.e. Commerce, OPIC, Export-Import Bank, and TDA. The 2017 Power Africa Annual Report estimates that in its four years, Power Africa has facilitated the financial close of 7,200MW of power generation capacity, helped to connect 10.6 million homes and businesses, engaged in 80 transactions worth $14.5 billion, and stimulated more than $500 million in U.S. exports.[29] OPIC, which originally pledged $1.5 billion to the effort, has more than exceeded this level of investment support. Funding in FY16 is estimated at $175 million from USAID, $643 million from MCC, $425 million from OPIC, and $14 million from TDA. (See Appendices).

Figure 4: Map of Africa Showing Power Africa Countries

Source: Free World Maps

29. USAID, Power Africa Annual Report, August 2017, available from *https://www.usaid.gov/sites/default/files/documents/1860/PA_FINAL_508c.PDF* .

Power Africa Funding: Key Countries in FY16 (N.B. Some Countries are funded from Regional Accounts)

USAID: Liberia, Congo (DRC), Mozambique, Ghana, Uganda, Djibouti, Tanzania, Nigeria, Ethiopia

MCC: Liberia, Tanzania, Ghana, Benin, Sierra Leone

OPIC: Ghana, Uganda, Guinea, Senegal, Nigeria, Tanzania

TDA: Ghana, Nigeria, S. Africa, Kenya, Sierra Leone, Zambia, Mozambique, Madagascar

With the creation by Congress and Secretary of State Clinton of the Energy Resources Bureau in the Department of State, the State established two global energy reform programs: one was on reform and improved management of oil and gas (Energy Governance and Capacity Initiative), including non-conventional gas, sectors in over a dozen developing countries; and a second focused on power sector improvements and investment issues (Power Sector Program) for example in Central America and the Caribbean. These programs were small (i.e. less than $10 million) compared with USAID energy assistance programs and sought to fill in gaps (e.g. in relationship to shale gas development experiences and issues) and be catalytic in nature. The Trump Administration requested about $6 million for the Energy Resources Bureau in its FY18 and FY19 budget requests.

One of the most important energy sector reform efforts that the United States government has supported is with Mexico and the fundamental energy policy and market reforms pursued by the Pena Nieto Government. On the electricity market side, both State and USAID have funded advisors that have been instrumental in developing the new competitive market framework and designing the renewable energy auctions that have led to enormous private sector investment in wind and solar technologies at prices that have been among the lowest in the world. The USDOE was also substantially engaged in energy cooperation activities with Mexico related to both cross-border natural gas and electricity trade issues and clean energy areas like energy efficiency. It appears that the Trump Administration has abruptly ended the USAID energy assistance activities in Mexico and no new energy obligations are reported in FY18.

During the first year of the Trump Administration, USAID, under its new Administrator Mark Green, who was formerly U.S. Ambassador in Tanzania and knows the Power Africa Program well, has begun to recast its energy programs and has clearly downplayed climate change. But USAID has continued to award new energy contracts in FY17 and FY18, including some dealing with renewable energy and energy access.

USAID released a major new solicitation in December 2017 that would contract with a group of U.S. companies that would be available to provide technical services for a broad scope of energy activities up to a total amount of $500 million. The objective of the work is stated as follows: "provide USAID Mission, Regional Bureau, host country stakeholders and partners with ready access to technical assistance and capacity building to manage the legal, regulatory, policy, governance, operational, financial, economic and transactional component of improved energy sector performance and strengthened enabling environment for private sector investment."[30] Four main cross-cutting themes are identified:

- Clean Energy: USAID supports countries to integrate and scale clean energy technologies.

30. USAID Request for Proposals (RFP) SOL-OAA-17-000108 Energy IDIQ II, December 14, 2017, p. 13, available from *www.fedbizopps.gov*.

- Energy Poverty: USAID supports countries in their efforts to expand access, while ensuring that the electricity supply is stable, resilient, affordable, and environmentally sustainable.
- Energy Sector Governance and Reform: USAID supports reforming and restructuring the energy sector and improving the accountability and transparency of authorities and institutions in the energy sector, as the foundations for improved performance and greater investment.
- Energy Security: USAID supports countries in their efforts to increase their energy security, through measures that encourage diversification of energy types and supply sources, promote use of sustainable domestic resources (such as renewable energy), reduce overall demand through energy efficiency and demand response measures, and facilitate better functioning of energy markets and greater integration through cross-border trade.

These prospective contracts replace a set of former energy contracts and will provide a convenient set of pre-negotiated instruments under which USAID missions and Bureaus can, if they wish, develop Task Orders for individual project implementation. The key question will be what budgetary resources will be devoted to energy programs as discussed in the Budget section below. As of this writing, no awards under this solicitation have been made.

V. Energy Technology Cooperation

This section addresses the role of energy technology cooperation in our overseas energy activities. Countries are interested in a wide-range of energy technologies in which U.S. national laboratories and private companies are engaged. Bilateral and multilateral activities are carried out, centered in the U.S. Department of Energy but also involving other agencies such as the Environmental Protection Agency and the Department of the Interior. Cooperation in this area has spanned the technology gamut from renewable energy and energy efficiency to clean coal and shale gas development. DOE's international program budget in FY16 and FY17 was about $18 million.

Although U.S. energy reform assistance programs contain a significant focus on technical issues, e.g. integration of renewable energy into electrical grids, there is a separate set of bilateral and multilateral technology cooperation programs.

The bilateral energy technology programs have tended to focus on major emerging market countries that are central to U.S. global energy policy concerns: i.e. China, India, Mexico, and Saudi Arabia. DOE is generally the lead for these interactions, often with funding from the OES (Oceans, International Environment, and Scientific Affairs) Bureau of the State Department. Under the Obama Administration, they were oriented towards renewable energy development and energy efficiency issues in support of U.S. global climate diplomacy. Programs with China (US-China Clean Energy Center) and India (Program to Accelerate Clean Energy – PACE) were the largest efforts and were managed by the DOE laboratories. Reported State funding for the US-China Clean Energy Center was $1.5 million in FY16.

The main multilateral efforts were with the Clean Energy Ministerial (CEM) process under the Major Economies Forum, the R&D Committee of the International Energy Agency, and the International Renewable Energy Agency, whose creation in Abu Dhabi the U.S. supported as an exception to our general policy on new international organizations. Other initiatives, such as the DOE-inspired Carbon Sequestration Leadership Forum and the Global Methane initiative led by EPA, were generally smaller in size.

The CEM, for which the U.S. served with State funding (e.g. $2 million in FY16) as the Secretariat through 2016, when the International Energy Agency took over the coordination role, established a set of technology-oriented working groups to share information and stimulate government and private investment. From this process arose the Mission Innovation Initiative that was developed for the Paris Climate Summit, involving 22 countries that pledged to double their clean energy R&D funding over five years to drive down costs and create new options to address climate change. It was paralleled by a private sector Breakthrough Energy Coalition championed by Bill Gates and other industrial leaders from various countries.

Leaders in the Mission Innovation Initiative agreed on seven major innovation challenges and selected lead countries for each:

1. Smart Grids (China, India, and Italy)

2. Off-Grid Access to Electricity (France and India)

3. Carbon Capture (Saudi Arabia and the United States)

4. Sustainable Biofuels (Brazil, Canada, China and India)

5. Converting Sunlight (EC and Germany)

6. Clean Energy Materials (Canada, Mexico and the United States)

7. Affordable Heating and Cooling of Buildings (EC, UAE and UK).

The United States has also played an important role in the emergence of the International Renewable Energy Agency, established in Abu Dhabi in January 2009. The US, through the State Department's International Organization Bureau budget, has been the third largest contributor to IRENA, after Abu Dhabi and

Germany, at about $4-5 million per year (e.g. $4.5 million obligated in FY16). IRENA has rapidly grown in membership to 154 members with 26 in accession status. It has become an authoritative source of information and advice on renewable technology and investments. Its products and services include: annual reviews of renewable energy employment, renewable energy capacity statistics, renewable energy cost studies, renewable energy readiness assessments, the global renewable atlas, renewable energy benefits studies, the Remap country analyses of renewable energy prospects to 2030, renewable energy technology briefs, facilitation of regional renewable energy planning, and renewable energy project development tools.[31] With the proposed budget cuts for the State Department and the U.S. participation in international organizations, the U.S. role in IRENA may decline.

The International Energy Agency, based in Paris and to which the U.S. contributed $618,000 in FY16, has had an active program of technology cooperation for many years. Focusing on renewable energy, energy efficiency, fossil fuels, and fusion, the IEA coordinates 39 Technology Partnership Agreements involving 53 countries.

VI. Energy Project Capital Assistance

Although USAID has provided substantial capital assistance for energy infrastructure over the years (e.g. rural electrification in Asia and power plants in Egypt, Pakistan, and Iraq) the focus of this type of assistance has narrowed to a smaller set of strategic countries. Formerly provided as loans, these projects are today financed on a grant basis generally under ESF account funding.

---

31. About IRENA, www.irena.org

While capital funding for power sector work in Iraq has been phased out, projects continue in Afghanistan and Pakistan. Afghanistan's efforts have focused on transmission and distribution infrastructure as well as hydro and solar generation. Reported funding in FY16 was $63.5 million including $5 million for the Kajaki Dam. The USAID Pakistan program has provided significant funding for rehabilitation of major hydro and thermal generation units, some of which the U.S. funded initially like the Tarbela Dam and the Guddu Combined Cycle Gas complex, as well as transmission projects. Funding for Tarbela Dam rehabilitation and wind transmission projects in FY16 were $25 million and $43 million respectively. The advancement of wind development in Pakistan is an interesting example of U.S. inter-agency cooperation and the combined use of technical assistance, capital assistance, and trade and investment promotion tools. While USAID provided technical assistance for energy regulatory development for renewables and planning for Pakistan's transmission grid as well as capital assistance financing for transmission infrastructure to integrate the planned wind farms, OPIC provided financing for U.S. developers, e.g. General Electric, of several wind farm projects in the Sind Province that were competing with Chinese and companies from other countries.

MCC has taken the lead role with respect to grant capital funding for energy infrastructure with numerous Compact and Threshold Agreements financing power generation, transmission, and distribution facilities. Over its 13 years, MCC has moved to incorporate significant policy and institutional reform conditions into its agreements to ensure the sustainability of projects and encourage private sector investment and has increased the level of its technical assistance to advance the reforms. Early energy compacts were in Georgia for gas pipeline rehabilitation and in Indonesia for small renewable energy investments. With the advent of the Presidential Power Africa Initiative, MCC has worked to develop Compacts in Africa for power sector development, including Ghana II ($498 million); Liberia ($257 million); Malawi ($351 million); Benin ($375 million); and Sierra Leone ($23 million), not including the earlier 2008 Compact for transmission in Tanzania. Following the earthquake in Nepal that damaged their power infrastructure, MCC developed a $500 million Compact that was signed in September 2017. A new Threshold agreement in Kosovo for $49 million will finance a Reliable Energy Landscape project including energy efficiency. MCC obligated $651 million for energy in FY16 (see Table below) out of total MCC obligations of $963 million, or over two-thirds.

Figure 5: Millennium Challenge Corporation FY16 Energy Funding

Country	Obligated Funding
Ghana	$417,313,578
Liberia	$201,080,058
Sierra Leone	$ 23,185,000
Nepal	$ 8,396,339
Benin	$ 906,464
Tanzania	$ 249,303
Total	$651,130,742

Source: USAID Explorer Dashboard, (https://explorer.usaid.gov)

Countries are happy to receive grant funding for energy infrastructure projects that otherwise they would probably have to be financed on a loan basis. They also see such projects as providing tangible illustrations of how foreign assistance is helping their economies. But given the limited U.S. funding in this sector, there is a strong case to be made that investments in institutional and human resource development may have greater and longer-term impacts for sustainable development. For instance, it is not unusual to see officials trained under USAID programs in the U.S. that are now leading their energy agencies. The U.S. has a comparative advantage in this sphere especially compared to China or other countries that focus on physical infrastructure and providing contracts for their companies. On the other hand, grant funding for capital projects like those from MCC can sometimes give the U.S. greater ability to influence recipient energy policies and institutional and governance reforms as well as advance U.S. commercial interests in these markets.

VII. Energy Trade and Investment Promotion

The so-called USG "trade agencies" have been a key component of the overall U.S. international energy cooperation effort. This section examines the roles of the U.S. Overseas Private Investment Corporation, the Trade and Development Agency, and the Export-Import Bank in promoting U.S. energy trade and investment overseas. With the focus of the Trump Administration on trade and renegotiation of agreements with major partners, this section will consider what changes are occurring and may occur in energy activities of these organizations.

The Trump Administration initially questioned the need for OPIC and TDA and President Trump's nominee to head of Export-Import Bank, who was on record as opposing its activities, was rejected by the Senate and sent back to the Administration. The budget proposal for FY18 proposed cutting funding for both OPIC and TDA. But the

Congress passed and the President signed the Build Act in October 2018, creating the U.S. International Development Finance Corporation (USDFC), expanding OPIC authorities and loan guarantee ceiling to $60 billion, and merging USAID Development Credit Authority loan guarantee program. The FY19 appropriations included funding for the transition to the new structure, which is expected to become operational at the beginning of FY20 (October 1, 2019).

OPIC has been very active in supporting overseas energy project investment, through both its insurance products and direct financing. Although, until recently, it was limited by Congress in its support for oil and gas development, it has moved significantly to ramp up activities in renewable energy. These reached $1 billion or more per year during the second Obama term. In 2016, OPIC reported energy activities of $1.4 billion, of which $55 million was for off-grid energy projects. OPIC has been a major player in the Power Africa Initiative, pursuing insurance and financing of projects, especially solar, wind, and hydro, in Nigeria, Guinea, Uganda, Senegal, Tanzania, and Ghana. OPIC has supported renewable energy project development in South Asia (i.e. wind in Pakistan and solar in India) and Central America and the Caribbean (i.e., solar in El Salvador and solar and wind in Jamaica).

A partner in the promotion of U.S. trade and investment has been TDA, a small agency with a budget of about $57 million in 2018 that helps catalyze U.S. exports of goods and services. Energy and infrastructure have traditionally been a strong focus of TDA and reported funding obligations for energy in FY16 totaled $30 million allocated to 75 activities in 26 countries. Although known for its support of pre-feasibility and feasibility studies that help U.S. companies compete with other country entities for larger infrastructure or power projects, TDA also pursues exchange activities and trade missions, often together with the U.S.

Departments of Commerce and Energy, that expose overseas officials to U.S. industries, companies, and technologies. TDA has also focused on renewable energy development working with Panama, Madagascar, South Africa, Ghana, Zambia, Vietnam, Jamaica, India, Belize, Tanzania, the Philippines, and others over the past two years.

TDA is also expanding its support for natural gas projects and U.S. exports and is leading a new U.S. Gas Infrastructure Export Initiative that was launched on November 17, 2017, in support of the Trump Administration policies to expand U.S. LNG exports. Under this Initiative TDA and the U.S. gas industry, including partners LNG Allies and the U.S. LNG Association, will work to advance gas infrastructure in key LNG-receiving countries through reverse trade missions, feasibility studies, training, and technical assistance. TDA's Annual Report for FY2018 shows energy activities remaining fairly constant at $30 million or over 50% of TDA's total $57 million in obligations.

The Export-Import Bank has historically been an important supporter for the export of U.S. energy technology and services. But it has had a troubled existence over the past few years, with a major debate over its extension and future in Congress. Although Congress did finally vote to extend its charter, it has been hamstrung since 2015 by the lack of a quorum of Directors on the Board which resulted in limiting its loan and loan guarantee authority to $10 million per project. Nevertheless, in FY17, Export-Import managed to underwrite $3.4 billion in loan guarantees. Export-Import's outstanding portfolio as of late 2017 was $72 billion. Both Power and Oil and Gas together have accounted for roughly 20% of this portfolio; as of FY17, power was about $4.195 billion (5.8%) and oil and gas represented 15.6% at $11.270 billion. The Trump Administration resubmitted to the Senate in January 2019 nominations for the President and three Board members of the

Export-Import Bank, but the full Senate has not yet taken up the nominations.

VIII. U.S. Role Vis-à-vis Other Countries and International Organizations

U.S. international energy cooperation activities do not occur in a vacuum. The way the United States government interacts with other countries and international organizations is critically important to understanding the political, diplomatic, and economic/commercial context and impact of the programs. This section will touch on some of the important issues in how the U.S. deals with other bilateral donors like China and such key international organizations as the World Bank and regional development banks.

The financing and investment requirements of the developing countries are huge and the amount of governmental and international financial institution financing is inadequate. Mobilizing private financing is critical. The Sustainable Energy for All Finance Committee completed a study in 2015 that tried to estimate funding levels to meet the 2030 goals of universal access, doubling of the rate of energy efficiency, and the doubling of the share of renewables. The report estimates that for the period from 2010-2030, more than $1 trillion per year is needed, broken down as follows: $49.4 billion for energy access (compared with about $9 billion currently); $442-650 million for renewables (from a base of about $258 million); and $560 billion for energy efficiency (compared with a current estimated $130 billion).[32]

The SE4All report calls not only for improvements in energy sector governance and management to reduce risks for investors but also

---

32. Sustainable Energy for All, SE4All Advisory Board's Finance Committee Report, July 4, 2015, p.1.

the development of financing mechanisms that can facilitate private capital flows and public-private partnerships, i.e. project preparation facilities, green bond market development, IFI de-risking instruments, insurance products focusing on removing specific risks, and aggregation structures that promote bundling and pooling approaches for small-scale opportunities.

The IEA's WEO 2017 estimates investment needed for power generation and transmission by region. Out of a world total of $6.741 trillion from 2017-2025, about $4.241 trillion is needed in the non-OECD countries, or about $471 billion per year over this period. [33]

The U.S. sometimes cooperates with other bilateral donors, international organizations, and international financial institutes to enhance and leverage its funds. Especially important are the World Bank and the regional development banks that are involved in policy dialogues and technical assistance as well as capital project loan financing. The U.S. Treasury, which represents the USG on the boards of these institutions, has been an important contributor to World Bank facilities such as the Global Environmental Facility and the Clean Technology Fund. Treasury was also the lead U.S. government agency on the Board of the Green Climate Fund, and the Obama Administration provided $1 billion of a $3 billion pledge prior to the reneging by the Trump Administration on the remaining amount. The GEF has allocated about $500 million toward clean energy and energy efficiency in its current 7th Replenishment. The overall enacted level of funding for Treasury support to GEF in FY17 was $146 million.

USAID and Treasury have worked closely with both the World Bank and the African Development Bank on the Power Africa effort that helped catalyze both institutions to expand their lending for the

---

33. IEA, World Energy Outlook 2017, Table 6.5, p. 252.

region. USAID obligated $3 million to the African Development Bank in FY16 for the AFDB's Sustainable Energy Fund.

Under AFDB President Donald Kaberuka, energy has become a top priority with a goal of allocating $3 billion over 5 years. AFDB is financing power transmission (e.g. Ethiopia) as well as generation projects in solar (e.g. Morocco, Mali); wind (e.g. South Africa and Kenya); hydro (e.g. Zambia, Tanzania, Nigeria); and geothermal energy (e.g. Kenya). The AFDB, like the World Bank, is using partial risk guarantees for some of their projects, given the poor financial condition of the electric utilities in most African countries. [34]

World Bank Group is a major source of financing for both technical assistance and energy capital projects. It reports energy financing of $49 billion since 2010 of which energy efficiency and renewable energy projects have assumed an increasing role with $21 billion. Energy financing in FY15 was $6.5 billion. The World Bank, IFC, and MIGA all seek to mobilize private and commercial financing. They have set a goal of $25 billion in commercial funds mobilized over the FY16-FY20 period.[35]

The Inter American Development Bank also has a considerable program in the energy sector, with a major emphasis on clean energy from renewables and natural gas. IDB funding of over $600 million since 2016 has included loans for solar in Mexico and Chile, geothermal in Nicaragua, wind in Argentina and Uruguay, energy efficiency in Jamaica and Mexico, and LNG to power in Brazil and Panama, as well as sector reform and rural electrification in Honduras and Bolivia.[36]

---

34. See www.afdb.org/energy and power.
35. See www.worldbank.org energy projects and programs.
36. See www.iadb.org/en.

The Asian Development Bank has a long history of lending in the energy sector, especially for hydropower projects. It currently has a broad mix of energy projects in its portfolio, including oil and gas and coal as well as renewables and transmission/distribution. In 2016-2017, major energy loans totaling over $4 billion per made, with Pakistan, India, Bangladesh, and Azerbaijan receiving the largest share.[37]

The expansion of Chinese overseas funding has been one of the most significant trends in the international development world over the past several years. China has become the largest source of energy finance with an estimated $43.2 billion in funding from the China Development Bank and the Export-Import Bank of China alone in 2016.[38] The Boston University China Energy Finance database estimates that over the period 2000-2016, the two Chinese banks provided $160 billion in energy finance (and up to $244 billion through 2018), mainly in Asia, concentrating on power plants projects in the coal and hydroelectric sectors. These efforts greatly exceed the combined energy loan commitments of the World Bank and regional development banks. The Chinese Belt Road Initiative has further intensified the Chinese international power and infrastructure push, with China pledging over $1 trillion in infrastructure financing. The creation of the Asian Infrastructure Investment Bank is one vehicle China has established to support its Belt Road initiative. The AIIB to date is undertaking 11 energy projects out of 25 listed as approved. They support a range of interventions in power transmission and distribution, natural gas, solar and hydropower development, and coal replacement, and air

---

37. See www.adb.org/sector/energy.
38. Kevin P. Gallagher, "China Global Energy Finance: A New Interactive Database," Boston University Pardee School of Global Studies, Global Economic Governance Initiative, www.bu.edu.

quality improvement in Pakistan, Bangladesh, Myanmar, Azerbaijan, India, Egypt and Tajikistan.[39]

The United States government has taken an increasingly negative view of the Belt Road Initiative and did not send high-level officials to the Belt Road Heads of State Summit last year in Beijing or the recent April 2019 meeting. U.S. officials have expressed concern about the impact of massive Chinese funding (termed "predatory financing"), such as in Pakistan or Sri Lanka, on the debt position of these BRI countries. These efforts are seen as undercutting the efforts of the IMF and international financing institutions to promote fiscal responsibility and transparency. This challenge from China certainly influenced the U.S. formulation of the "Free and Open Indo-Pacific Strategy," announced during President's Trump visit to Vietnam for the Asia-Pacific Economic Summit in November 2017,[40] and its inclusion of a market oriented emphasis on energy and infrastructure development and cooperation with Australia, India and Japan as well as ASEAN nations. During the Obama Administration, there was considerable debate over how best to deal with the growing Chinese overseas financing. Former Obama USAID Administrator Raj Shah was a strong proponent of cooperation with China and discussed with the Chinese cooperation on development of the gigantic Inga Hydro Project in the DRC and other possibilities in the context of the Power Africa Initiative. His successor, Acting USAID Administrator Lenhardt, signed a Memorandum of Understanding with the Chinese Ministry of Commerce in 2015.[41] In the past, some USG officials have sought

---

39. Approved Projects Overview – AIIB, www.aiib.org.
40. White House, "Remarks by President Trump at APEC CEO Summit, Da Nang, Vietnam," November 10, 2017, available from https://www.whitehouse.gov/briefings-statements/remarks-president-trump-apec-ceo-summit-da-nang-vietnam/
41. "Memorandum of Understanding on US-China Development Cooperation and the Establishment of an Exchange and Communication Mechanism, September 2015, www. usaid.gov.

to encourage greater Chinese collaboration and co-financing with the IFIs to bring more transparency to Chinese projects and ensure that proper standards and concerns for the governance, corruption and environmental implications of development financing were applied. Current USAID Administrator Mark Green has been critical of the BRI initiative and cautioned countries about being caught in a "debt trap." In his April 26, 2019 address to the Belt Road Summit, President Xi tried to respond to growing criticisms by proposing a new debt-sustainability framework and pledging greater transparency.[42]

IX. Congress and the Budget Process

Although international energy program funding has not been a big slice of the past foreign operations budgets and expenditures (compared to health for example), it has not been insignificant. The Trump Administration's shifts on energy and environment, however, raise uncertainty over future funding for this sector. The preliminary numbers seem to suggest a reduction in levels in FY18 to about $1.2 billion, a drop from $2.5 billion in FY16 and $2.1 billion in FY17, with the greatest fall in MCC to around $50 million and OPIC to $872 million from $1.35 billion. These shifts are occurring within the broader context of the Trump Administration's efforts to make deep cuts in the Foreign Operations Account, i.e. 31% proposed cut in International Affairs in FY19 or $41.7 billion.

Strong voices in Congress have expressed opposition to these cuts in development and diplomacy funding. The Executive Branch has operated under Continuing Resolutions in recent years and Congress passed on March 22, 2018, a FY18 Omnibus appropriations bill within the two-year budget framework approved in early February

42. Chun Han Wong and James T. Areddy, "Xi Signals Changes for "Belt and Road", The Wall Street Journal, April 26, 2019, p. A8.

2018. The enacted foreign operations budget under the FY17 Omnibus Act contained $8.7 billion for health, $2.995 for Development Assistance, and $1 billion for ESF, and $291 million for Europe and Eurasia, $905 million for MCC, and $75 million for TDA. Treasury was appropriated $146 million for the GEF (Global Environmental Facility) and $1.3 billion for World Bank IDA.[43] The Senate Foreign Operations Committee Report recommended $291 million for Power Africa and $177 million for clean energy in the DA budget and $2 million for the Caribbean Energy Security Initiative

The Foreign Operations budget submission for FY18 omitted climate change from its presentation and recast most country program descriptions to eliminate line-items for energy. The line item on environment in the FY18 budget request was reduced from $1.7 billion to $185 million, not including funding for the GEF. The proposed FY18 budget merged development assistance and ESF accounts into one Economic Support and Development Fund with a request level of $4.9 billion which represented a large $4.2 billion decrease in the merged budget. The Congress rejected this approach in the final FY18 Omnibus, actually increasing funding for OPIC and TDA and keeping DA funding flat at $3 million while reducing funding for ESF ($713 million or 15.2%) and AEECA ($168 million or $16.8%).[44] The Administration's FY19 request again proposed cuts of 22.7% below the FY2018 enacted level. But in the agreement following the government shutdown, the final bill appropriated $54.376 billion or slightly more than the FY18 level.[45]

---

43. "State-Foreign Operations Appropriations, Fiscal 2017: Where the Money Goes" www.media.cq.com.
44. US Global Leadership Coalition, "Congress Finalizes FY18 Spending: Rejects Doctrine of Retreat, Restores Funding for International Affairs Budget," March 23, 2918, www.usglc.org.
45. Congressional Research Service, "Department of State, Foreign Operations and Related Programs: FY2019 Budget and Appropriations," updated March 12, 2019, available from https://fas.org/sgp/crs/row/R45168.pdf, p.2.

Based on the obligations data and despite the reduced overall energy funding levels, there does not appear to have been a huge shift in the country's focus of programs. Pakistan, Ukraine, Nigeria, and India still occupy prominent positions. Power Africa funding has definitely been reduced, but country programs and OPIC investments in Africa, particularly in renewable energy, have continued. Specific obligations were untaken in FY18 in Liberia, Ethiopia, and a number of other Sub-Saharan countries.

Besides the major priority of assistance to Ukraine, support for countries in the Balkans and the Black Sea continued, notably in Kosovo, Bosnia, Albania, Georgia, and Armenia. Except for India, Nepal, and Vietnam, the levels for Asia did not yet increase in FY18 with the Asia EDGE initiative but may do so in FY19 and FY20. Columbia remains a focus in South America and continued support for the Caribbean in the wake of the Venezuela crisis is likely.

X. Conclusions

Energy issues are of growing importance to U.S. foreign policy and the expanding markets in the developing, non-OECD countries are a key factor influencing U.S. international energy policy and programs. Under the Obama and Trump Administrations, the United States government has pursued a range of international energy assistance and cooperation activities with developing countries to advance U.S. security, trade, development, and environmental interests.

Looking back at Title V of the Nuclear Nonproliferation Act of 1978, its main theme of cooperation with developing countries in helping them meet their energy needs through renewable and non-nuclear alternatives has been and is being pursued by various agencies of the U.S. government. Title V's proposed energy sector and resources assessments, trade facilitation, partnerships, training, technology cooperation, and technical exchanges have become a centerpiece of U.S. overseas energy work.

Under the Obama Administration, international energy programs focused on "clean energy" and were largely grouped under the climate change umbrella and the GCCI (Global Climate Change Initiative), with the exception of Power Africa and ESF and AEECA programs in such countries as Pakistan, Afghanistan, Jordan, Ukraine, and Georgia. During this period both OPIC and MCC increased substantially their focus on energy investment and sector development with MCC becoming the largest source of U.S. grant capital assistance. Overall U.S. funding for international energy assistance and cooperation (excluding Export-Import Bank guarantees) in FY16 is estimated at something over $2.5 billion. The main geographical focus of these programs and projects has been in Sub-Saharan Africa and the Near East (including Afghanistan and Pakistan).

The advent of the Trump Administration in 2017 saw a dramatic shift in orientation and view toward government funding for development assistance and U.S. international investment, with proposals for deep cuts in State and USAID and the elimination of both the Overseas Private Investment Corporation and the Trade and Development Agency. In energy, the new Administration embraced the notion of U.S. "energy dominance" and the use of U.S. abundant coal, gas, and oil resources and exports to further U.S. trade and foreign policy objectives. The Obama focus on international assistance and international multilateral cooperation to tackle climate change was eliminated, and the budget submissions for FY18 and FY19 generally removed or cut significantly line items for energy and environment. Although in the final FY18 and FY19 Appropriations Acts Congress resisted the proposed deep cuts in the International Affairs budget and the Trump Administration is attempting to develop the Asia EDGE energy initiative, actual funding levels for international energy programs have fallen and remain tenuous.

The Administration and Congress need to recognize that big cutbacks in international energy assistance and cooperation programs will hurt the United States' competitive position vis-à-vis other countries like China that are ramping up financing in their aggressive pursuit of developing country energy markets. While most developing countries in the post-Paris Climate framework are expanding their renewable energy and energy efficiency investments, the Trump Administration is emphasizing fossil fuels. A reduction in U.S. programs to support renewable energy would work against the market forces that are furthering the expansion of these sources due to favorable cost, technology, and environmental factors. Although U.S. LNG exports can help some countries diversify their energy mix, the United States needs to take advantage of its strong renewable technology base to help countries pursue these sources.

The experience of the Power Africa "whole of government" program, with its bipartisan Congressional support in the Electrify Africa Act, has generally been positive. Its strong private investment and transactional approach has worked and can serve as a model that can be adapted to U.S. energy assistance and cooperation in other regions.

The Congress and the Trump Administration need to create a new framework for international energy activities in the foreign operations budget that allows an "all of the above" energy approach to receive greater visibility and funding. The leadership of State, USAID, and DOE is critical to this process, but the key trade and investment roles of the new U.S. Development Finance Corporation, TDA, and Ex-Import Bank, as well as the Departments of Treasury and Commerce, in this strategy, need to be recognized and the agencies supported. Senate approval of the full Export-Import Bank Board and the restoration of the full functionality of the Export-Import Bank is critical to future U.S. international energy leadership and competition with China.

Appendix A: Estimated USG International Energy Funding

(FY16 and Preliminary FY17)

(Excluding Export-Import Bank and Treasury IFI)

(Dollars Millions)

Agency	FY16	FY17
USAID	446	289
MCC	651	386
OPIC	1400	1365
TDA	30	35
STATE	12	12
DOE	18	18
TOTAL	2557	2105

Sources: USAID Foreign Assistance Explorer (https://explorer.usaid.gov/#2018), TDA and OPIC Annual Reports and web sites; State number is a rough estimate.

Appendix B: USAID Energy and Related Environmental Assistance Summary

Obligations in FY16 and FY17 (millions)

Area	FY16	FY17
Global	20.9	16.1
Africa	170.5	57.2
Middle East & S.Asia	178.6	160.6
Other Asia	12.3	8.6
Latin America	23.0	22.9
Europe and Eurasia	19.6	23.8
Total	424.9	289.3

Source: USAID Foreign Assistance Explorer (https://explorer.usaid.gov/#2018)

Appendix B Detail: USAID ENERGY AND RELATED ENVIRONMENTAL ASSISTANC FY2016
Obligations by Region and Country ($ Millions)

	Country	Activity	Obligation
Africa			
	Regional	Power Africa Transaction and Reform	78
	Regional	East Africa Regional Construction	10.6
	Regional	Power Africa ElectriFi Agreement	10
	Regional	Power Africa:AfDB Sustainable Energy Fund	10
	Regional	Power Africa Investor One Facility	5
	Regional	Scaling Solar Partnership with World Bank	4.2
	Regional	Modern Energy Services	7.85
	Regional	Modern Energy Services	1.8
	Regional	Climate Economic Analysis	1.75
	Regional	USAID-Shell Foundation Coop. Agreement	7
	Regional	USEA Energy Utility Partnership	1.4
	Regional	Power Africa Delivery Unit	1.3
	Regional	Power Africa Senior Advisors	1
	Regional	Power Africa Transaction and Reform	0.9
	Regional	Enhancing Sustainable Utility Regulation	0.9
	Regional	Regulatory Partnership	0.8
	Regonal	Gender Equality for Climate Change	0.8
	Regional	Modern Energy Services	0.4
	Regional	Partnerships for Enhanced Research	0.3
	Regional	Power Africa Transaction and Reform	3.7
	Regional	AfDB	3
	Regional	Power Africa Transaction and Reform	0.5
	Regional	LEDS	0.3
	Regional	Financing Growth	0.1
	Congo (Kinshasa)	Power Sector Reform	5
	Mozambique	Support Policy Env. And Economic Deve.	2.5
	Ghana	Integrated Resources and Resilience Planning	1.9
	Ethiopia	Scaling Solar Partnership with World Bank	1.7
	Tanzania	Modern Energy Services	1
	Uganda	Modern Energy Services	1
	Uganda	Modern Energy Services	0.8
	Uganda	Cooperative Dev. Rural Electrification NRECA	0.8
	Uganda	Modern Energy Services	0.5
	Uganda	Modern Energy Services	0.3
	Uganda	Environment Management for the Oil Sector	2.9
	Dijbouti	Utility Partnership Program	0.4
	Dijbouti	Power Africa Transaction and Reform	0.8
	Nigeria	Modern Energy Services	0.3
	Madagascar	Scaling Solar Partnership with World Bank	0.1
	Zambia	Modern Energy Services	0.1
	Kenya	Powering Agriculture	2
	Kenya	LEDS	1.2
Subtotal Africa			174.9

Asia		
Bangladesh	Rural Electrification with World Bank	0.7
Bangladesh	Clean Energy	3.9
Philippines	LEDS	3.5
Indoneia	Energy Utility Partnership Program	0.1
Indonesia	Sector Reform and Utiity Commercialization	0.4
Indonesia	Clean Energy Development	5.1
China	Smart Cities (City Links)	2.5
Regional	LEDS	1.1
Regional	Clean Power	2.2
Regional	Private Financing Advisory Network	1.7
Regional	LEDS with DOE	0.3
Regional	EC-LEDS	2
Vietnam	Clean Energy	1.7
Nepal	Hydropower Development	1.5
Vietnam	LEDS	1.2
India	Grid Integration with DOE	2
India	PACE-D	2.4
India	Utility Partnership	0.5
India	Energy Utility Partnership Program	0.7
India	Enhancing Sustainable Utility Regulations	0.5
India	LEDS with DOE	0.3
Subtotal Asia		34.3
Near East		
Aghanistan	Engineering Support	60.4
Afghanistan	Kajaki Dam Hydropower Project	5
Afghanistan	Modern Energy Services	0.1
Pakistan	Guro Keti Bandar Wind Power Project	43
Pakistan	Tarbela Dam Repair and Maintenance 2	25
Pakistan	Clean Energy for Pakistan's Economy	10
Pakistan	DCA Bank Loan Portfolio Guarantee	7.5
Pakistan	Training for Pakistan	2.9
Pakistan	Agreement of Dept of Army for Develop Water	1
Pakistan	Commercial Dev. With Commerce Dept.	0.8
Pakistan	Modern Energy Services	0.7
Pakistan	Energy Policy Project	0.6
Pakistan	Utiity Partnership	0.2
West Bank/Gaza	Palestine Renewable Energy Project	5
Jordan	Energy Sector Capacity Building	6
Subtotal Near East		168.2

Latin America			
	Regional	Clean Energy Finance Facility	6
	Regional	Central America Energy Initiative	4.2
	Colombia	Clean Enegy	3.6
	Haiti	Sector Reform and Utility Commercialization	1.1
	Guatemala	LEDS	3
	Mexico	LEDS	3.2
	Mexico	LEDS	1.4
	Mexico	Utility Regulation	0.3
	Mexico	LEDS	0.2
	Jamaica	Sector Reform and Utility Commercialization	1.9
Subtotal Latin America			24.9

Europe and Eurasia			
	Ukraine	Municipal Energy Reform Project	7.5
	Ukraine	Financing Growth	1.7
	Ukraine	Regulatory Partnership	0.2
	Ukraine	Utility Partnership	0.2
	Regional	Eurasia CASA-1000	2
	Regional	USEA Partnership	1.5
	Regional	LEDS	1.5
	Regional	Energy Regulatory Partnership	1.5
	Regional	Central Asia Smart Water Project	0.7
	Georgia	EC LEDS Clean Energy Program	2.5
	Georgia	Energy Partnership USEA	0.4
	Georgia	Regulatory Partnership	0.1
	Macedonia	Modern Energy Services	0.3
	Kazakhstan	Climate Mitigation	1.8
Subtotal Europe and Eurasia			21.9

Global			
	World	Financing Growth	0.2
	World	Renewables, Energy Efficiency, Grid Manage.	3.6
	World	EC LEDS	2.3
	World	Powering Agriculture	10.9
	World	EC EDS	1
	World	Clean Energy Emissions Reduction	0.8
	World	Sector Reform and Utility Commercialization	0.8
	World	Energy Efficiency and Renewables	0.7
	World	Energy Regulatory Partnership NARUC	0.6
	World	Energy Utility Parternship USEA	0.6
	World	Energy Sector Technical Leadership	0.6
Subtotal Global			22.1
Total All			446.3

Source: USAID Foreign Assistance Explorer (https://explorer.usaid.gov/#2018.

Appendix C: Energy Funding by the Millennium Challenge Corporation (MCC)

FY16 and FY17 by Country ($ Millions)

Country	FY16	FY17
Ghana	418	0
Liberia	201	0
Sierra Leone	23	0
Benin	0.9	0
Tanzania	0.2	0
Senegal	0	263
Nepal	8.4	71
Total	651.5	334*

Sources: Annual Reports for 2016 and 2017. *USAID Foreign Assistance Explorer (https://explorer.usaid.gov/#2018) show $386 million for MCC obligations in FY18.

Appendix D: Funding for Energy (Finance and Insurance) by the U.S. Overseas Private Investment Corporation (OPIC) FY17 by Region – ($ Millions)

Region:	FY17
Africa	342.1 million (292 finance; 50 insurance)
ME and S. Asia	586.9 million (46.9 finance; 539.1 insurance)
Latin America	436.5 million (407.25 finance; 29.25 insurance)
Total	1365.5 million

Sources: The OPIC web site indicates that OPIC energy activities, both finance and insurance, totaled $1.4 billion in 2016. No breakout is given. Detailed data in the OPIC Annual Report for FY16 don't seem to add up to this much. Numbers for FY17 below are closer.

Appendix E: Energy Project Funding by the U.S. Trade and Development Agency

(Obligations for FY16 and FY17 by Region) ($Millions)

Region	FY16	FY17
Africa	14.793	22.175
ME and S. Asia	0.813	2.494
Other Asia	3.328	6.428
Latin America	8.122	5.501
Europe	0.380	0
Global	0.923	1.242
Total*	28.359	37.843
FY 2017 Annual Report:	**30.407**	**32.523**
Foreign Aid Explorer total:	30	35

Sources: *Total based on review of summary detailed obligations data from TDA Annual Reports and USAID Foreign Assistance Explorer (https://explorer.usaid.gov/#2018

Motivations for Nuclear Power Programs

Stephen Thomas

1. Introduction

Arguably, over its 60-year commercial history nuclear power has seldom, if ever, been the cheapest way to add generating capacity to an electricity system and there is little sign that nuclear power is becoming more competitive. Yet the enthusiasm of many utilities and governments for the nuclear option appears, in many cases, to be undimmed by the disappointing results of earlier programmes and some countries appear keen to launch new nuclear programmes. In this paper, we identify three key factors, weak economic regulation, availability of low-cost finance, and military aspirations that may allow or motivate the ordering of economically unjustifiable nuclear power plants, focusing mainly on weak regulation. We then review the four case study countries, China, Saudi Arabia, Iran, and Taiwan, to determine how far these factors apply in those countries.

Nuclear power is often justified by its promoters by optimistic forecasts of costs either from the utility or from the reactor vendor and with the complicity of government. Even when it is known that nuclear is not the cheapest option, it is justified by claims that ordering now would open a stream of cost reductions from effects such as learning, technical progress, and economies from mass production methods that would make nuclear power economic. A capability to offer nuclear power technology is also sometimes seen as one that can open export markets. It appears that, for example, the nuclear sectors in France, China, and Russia are supported by national government, and orders in the home market are encouraged to provide a shop window for the country's nuclear technologies.

The nuclear option is also claimed to have non-monetary strategic advantages; for example, diversifying fuel sources, reducing energy import dependence, and reducing pollution resulting from use of fossil fuels. However, while nuclear power might be seen to have advantages in these factors, it is seldom compared with other policy options that would achieve the same goals. It is consumers that almost invariably must pick up the financial consequences of investment in uneconomic plants. Despite this, any public resistance to nuclear programmes is seldom on the grounds of cost.

There would seem to be three factors that, often in combination, allow or motivate the ordering of uneconomic nuclear power plants:

- The plant buyer is insulated from the consequences of uneconomic investment decisions by weak regulation of electricity prices;
- Low-cost finance is available; and
- The plant contributes, either overtly or covertly, to creating or maintaining a nuclear military capability either for weapons or submarines.

2. Motivating Factors

The analysis in this paper focuses mainly on the first factor, weak regulation, but the other two factors are also examined.

2.1. Lack of Financial Responsibility

For most of its history, the electricity industry has been seen as a natural monopoly. Prices were generally set by the utilities and often publicly-owned, sometimes via some form of regulation, although in most countries this regulation was neither independent of government nor rigorous. Since around 1990, there have been increasing attempts to transform the generation part of the electricity business into a competitive market, under which new power plants would have to compete in a market to repay the investment. If competition was vigorous, this would mean that investment in new power plants of all types would become very risky. This would particularly impact capital intensive options because it would raise the cost of borrowing, and nuclear is the most capital-intensive form of generation. Because capital-intensive plants would still have to meet the high fixed costs associated with repaying the construction cost whether the plant was able to compete successfully or not.

2.1.1. Fully Regulated Jurisdictions

In the 1930s, the USA established and continues to operate a rigorous price-setting framework at state-level.[46] Essentially, prices are set to allow the utility to recover operating costs and make a 'fair' return on prudent utility investment. This requires regulators to determine when a new asset is complete and ready to enter service, whether 'the costs were prudently incurred,' and whether it will be 'used, useful' before the asset is added to the portfolio of investments on which the utility can make a return. If the investment does not meet these criteria, the utility will not be allowed to recover

46. Union of Concerned Scientists (undated) 'Public Utilities Holding Company Act' Union of Concerned Scientists, Washington. https://www.ucsusa.org/clean_energy/smart-energy-solutions/strengthen-policy/public-utility-holding.html (Accessed May 15, 2019)

the imprudent costs from consumers and will be required to pay for these costs from its profits.

Up until the mid-1970s, while the electricity industry was on a downward cost curve and real electricity prices were falling, this system appeared to work smoothly. In short, the decision for regulators was often seen as when to reduce prices and by how much. However, a combination of lower than expected electricity demand, rising fuel costs, and a high proportion of nuclear projects being completed very late and far over-budget put a heavy strain on the system. More than 120 firm reactor orders, including some which were claimed to be nearly complete, were cancelled because they would not have been judged as 'used and useful' and the costs were not 'prudently incurred.'[47] This abandonment of orders was often facilitated by the state regulators, allowing utilities to recover some of their costs only if the projects were cancelled. This meant the costs passed on to consumers were less than they would have been if the plant had been completed. It also reduced the risk to the utility that the cost of completing the plant would continue to escalate with little hope of cost recovery from consumers.

These actions quickly brought a halt to nuclear ordering in the USA because financiers quickly saw that nuclear orders were now very risky to the utility and were unwilling to lend to them. No orders were placed after 1979 and all orders placed after 1974 were cancelled until the two projects, Summer and Vogtle, that began construction in 2013. Summer was abandoned in 2017 and it is far from certain that Vogtle will be completed.

President G W Bush announced in 2002 a programme of Federal subsidies, under the Nuclear 2010 policy, for a small number of

---

47. Wikipedia (2019) 'List of cancelled nuclear reactors in the United States' Wikipedia.
https://en.wikipedia.org/wiki/List_of_cancelled_nuclear_reactors_in_the_United_States (Accessed May 16, 2019)

demonstration plants in the belief that these would lead to nuclear ordering without subsidy. By then, several states in the USA had abandoned the fully regulated model for the electricity industry in favour of a competitive model.

Utilities in both regulated and deregulated states quickly announced plans to build more than 30 new reactors under these provisions, but in the event only the Vogtle and Summer projects, both for two reactors, went forward. Both projects are in fully regulated states and they were possible because the state regulators gave strong indications that costs would be recoverable and allowed the utilities to begin to recover their costs from the start of construction, not the completion of the plant. Both projects went wrong from the start with costs and construction times escalating alarmingly culminating in the bankruptcy in 2017 of the vendor, Westinghouse (then owned by Toshiba of Japan). Westinghouse had signed effectively fixed price deals with the utilities and, when costs escalated, it could not pass them on to the utility customers. In summer 2017, the Summer project was abandoned and Vogtle appeared at severe risk of also being abandoned, although in December 2017, the Commissioners of the Georgia Public Service Commission (PSC) agreed, against the advice of the PSC staff, for the project to continue. In the Summer case, consumers will lose much of the money already paid to the utility for construction. It is still far from certain whether the Vogtle project will be completed, and if it has to be abandoned, consumers will also lose much of the money paid by them up until then.

Given this experience, there appears little prospect that regulators in other regulated U.S. states will give utilities favourable indications on cost recovery sufficient to allow nuclear projects to go ahead. Utilities and, more importantly, financiers will see nuclear projects as far too risky to contemplate.

2.1.2. Competitive Electricity Markets

Outside the USA, utilities were able to build nuclear power plants with the knowledge that whatever costs were incurred could be recovered from consumers. This allowed utilities to make investment decisions for which they were insulated from the risk. It was the perception of a 'cost-plus' culture that led to the interest in competitive electricity markets under which it was assumed utilities would be forced to seek the lowest cost ways to add generating capacity. If they made uneconomic investment decisions, they would be forced to sell their power at a loss and perhaps go bankrupt. All the European Union (EU) member states are now required to operate their electricity systems on a competitive basis with wholesale power markets and full retail competition for all consumers mandatory.[48]

This trend of competition was widely seen as being very bad for nuclear power because of its capital intensity, the high cost of nuclear generation, and because the high likelihood of cost and time overruns for nuclear projects made them very risky. This expectation of a bleak outlook for nuclear power in the EU has proved well-founded with no nuclear plants ordered and completed since electricity liberalisation was required in 1997. Only four member states have reactors under construction or at an advanced stage of planning: Finland, the UK, France, and Hungary. All of these except France have had to take measures to fully insulate the reactors from the market and all the projects appear problematic.

Finland has one plant under construction and expects construction on a second, which has been delayed by several years, to start in

48. Liberalisation is required under EU Directive 2009/72/EC. Official Journal of the European Union (2009) 'Directive 2009/72/EC of the European Parliament and of the Council' European Commission, Brussels. https://eur-lex.europa.eu/legal-content/EN/TXT/PDF/?uri=CELEX:32009L0072&from=en (Accessed May 16, 2019)

2018/19.[49] As has been well documented, the first, Olkiluoto, starting construction in 2005, has gone badly wrong and is now 11 years late and about three times over-budget.[50] The utility, TVO, thought the risk would be mitigated by two main factors: first, the plant was ordered under a fixed price (turnkey) contract for €3bn; and second, most of the output was contracted to be sold at cost to the consortium of electric-intensive industrial consumers that owned the utility. Neither measure has worked. The vendor, Areva, denied responsibility for a large part of the additional costs and this dispute is being settled in the Court of Arbitration. It seems inevitable that TVO will have to bear a significant part of the cost overruns. This will make power from the plant very expensive and, given that electric-intensive industries can only survive if they have access to cheap power, it is hard to see how they can justify buying power from this plant when power from the competitive market, NordPool, is so much cheaper. If its customers are unwilling to buy power from Olkiluoto, the future of TVO will be in doubt. Despite this experience a second plant is planned, purchased from Rosatom (Russia), and is at an advanced state of planning under similar arrangements but with Rosatom taking a minority equity stake.

The UK launched a new effort to build reactors in 2006 but with a firm promise that no public subsidies would be given. This promise was essential to making the programme politically feasible and was to deal with fears that, as with previous nuclear programmes, large extra costs would accumulate and would fall on consumers and taxpayers. The clear implication of the no-subsidies policy was that new nuclear plants would compete in Britain's wholesale electricity market on equal terms with other generations. Under this programme, 16GW of new capacity (11 reactors at five sites) is

---

49. Nuclear Intelligence Weekly 'Paper Chase Slows Finland's Hanhikivi Project to a Crawl' April 26, 2019, p 5.
50. Nuclear Intelligence Weekly 'EPR Teething problems continue' April 19, 2019, p 3.

planned. More than a year before construction on the first reactor is expected to start, power from the first reactor is at least 10 years late and the expected cost per reactor has increased five-fold.[51] The no-subsidies promise was never credible, and a series of large public subsidies have had to be given to prevent the programme collapsing, although the no-subsidies promise was not formally abandoned until 2015. Subsidies include a cap on waste disposal costs, a 35-year fixed real price take-or-pay contract to buy the output at a cost far above current wholesale prices, and loan guarantees to cover all the borrowing required. Despite all these measures, construction start on the first reactor was only in December 2018, three of the five projects have been abandoned and a fourth is in doubt.[52]

Hungary is expecting to start construction of two reactors bought from Rosatom after 2019 and the output of these will be bought under long-term contract. The loan deal has already been signed and Hungary will have to start repayments to Russia in 2026 whether or not the reactors are in service.[53]

The French electricity market is theoretically open to competition but the incumbent, Electricité de France (EDF), still has about 85 percent of the market.[54] This allowed it to start construction of the Flamanville plant in 2007 with no measures to insulate the plant from the market. As with its sister Olkiluoto plant, construction has gone badly wrong and it is now at least nine years late and far over

---

51. S D Thomas (2017) 'Time to cancel Hinkley?' Stop Hinkley, Bridgwater. http://www.no2nuclearpower.org.uk/wp/wp-content/uploads/2017/09/Time-to-Cancel-HinkleyFinal.pdf (Accessed May 16, 2019)
52. Nuclear Intelligence Weekly 'Hitachi UK Exit Signals Industry Turning Point' January 11, 2019, p 1
53. Nuclear Intelligence Weekly 'Five Years on, Hungary's Paks Expansion Stumbles Along' February 18, 2019, p 6
54. Reuters 'French alternative power providers grab more market share in Q2 – CRE' September 29, 2017. https://af.reuters.com/article/energyOilNews/idAFL8N1MA59B (Accessed May 16, 2019)

budget. Plans to build follow-on plants did not materialise and in 2017, EDF stated that it could not build reactors in France without 'guaranteed income.'[55]

In the USA, nuclear projects were planned in states with competitive markets, for example, in Texas and Maryland. The Calvert Cliffs (Maryland) and the South Texas projects were short-listed for loan guarantees. The state regulatory body of Maryland was unwilling to give any guarantees and the project collapsed. The fee for the loan guarantees, reported to be as high as 11.6 percent because of the project risk, was also a factor in the project collapse.[56] The South Texas project collapsed before it got to the stage of negotiating loan guarantees.

2.1.3. Vendor Price Guarantees

Vendors very seldom sign fixed price (turnkey) contracts that would expose them to any cost overruns. Where they do, the consequences to them are often financially very serious. This was the case with the fixed price contract given by Areva for €3bn for the Olkiluoto plant in 2004. By 2018, the plant was still at least a year away from commercial operation, the costs had escalated three-fold or more and this was a significant factor in the financial collapse of the vendor Areva NP.[57] Westinghouse became exposed to US$6bn of the

55. Reuters 'MEDIA-EDF says no new nuclear reactors in France without state support-Ouest France' December 15, 2017.
https://www.reuters.com/article/edf-nuclearpower/media-edf-says-no-new-nuclear-reactors-in-france-without-state-support-ouest-france-idUSL8N1OF3EP (Accessed December 20, 2017).
56. Michael J. Wallace, Chairman of Constellation Energy Nuclear Group, letter to Dan Poneman, Department of Energy, 8 October 2010.
57. Nuclear Intelligence Weekly 'End In Sight After Settlement on €11.4 billion-Plus OL3' March 16, 2018, p 3.

cost escalation at the Vogtle and Summer projects and this was the major factor in its bankruptcy in 2017.[58]

2.2. Availability of Low-Cost Finance

Under the old monopoly, regulated market model, the financial risk of nuclear projects was almost entirely borne by consumers rather than utilities and so lending to a utility was extremely low risk - utilities were the original 'blue chip' investment. Also, most utilities outside the USA were publicly owned and there was an element of trust in government that such companies would not engage in the profiteering that might be expected of investor-owned companies. In some cases, notably, Electricité de France (EDF) until it was part-privatised in 2005, the government guaranteed all EDF's debts so effectively lending to EDF was no riskier than lending to the French government. So EDF's credit rating was as high as that of the French government, AAA, and the cost of borrowing correspondingly low. This very cheap finance was an important element in allowing EDF to finance the nuclear programme of the late 70s and early 80s when about 40 reactors were ordered in about six years.

Some countries had too low a credit rating for their utilities to obtain the finance needed and, for example, attempts to launch nuclear power programmes around 1980 in countries such as Turkey and Egypt were unsuccessful because they could not be financed.

As the real cost of nuclear power plants has continued to escalate and the pressure on utilities to make only economic investments has increased, the availability of cheap finance has increasingly become the determining factor as to whether a nuclear programme is feasible. In a few countries with large financial resources and a

58. Nuclear Intelligence Weekly 'Toshiba Cuts Ties With Westinghouse' March 31, 2107, p 3.

government policy to promote nuclear power, notably China and, in the future, Saudi Arabia, finance is likely to be available. However, in the past two decades, in most other countries, nuclear programmes are only viable if finance is offered by the vendor or a government, either that of the vendor or that of the purchaser, is willing to offer sovereign loan guarantees.

The only vendors able to offer finance are those backed by a strong government in a centrally planned economy, i.e. the Russian and Chinese vendors. Rosatom (Russia) has a large reactor export order book, almost all of which will require Russia to supply the finance at relatively low-cost. The Chinese vendors (CGN, CNNC, and SNPTC) have yet to win reactor export orders (except for Pakistan) but in the markets that they are targeting, finance is reported to be on offer from Chinese state-owned banks.

Sovereign loan guarantees provided in 2004, mainly by the French government, were used to help finance the Finnish Olkiluoto project supplied by the French state-controlled vendor, Areva NP. These guarantees were worth €700m, about a quarter of the expected cost, and their availability was enough to persuade a consortium of banks to offer the finance required. The construction problems with many of the nuclear construction projects have strengthened the financial community's perception of nuclear as a highly risky technology. This, and the increased real cost of nuclear plants, has dramatically increased the scale and level of cover from loan guarantees required. For example, when the Bush Nuclear 2010 policy was conceived in 2002 to support about five new reactors, it was expected that nuclear plants would cost about US$1000/kW and that 40 percent of the construction cost would be covered by guarantees. This would have required guarantees of about $3bn (assuming 1500MW reactors). A decade later, the expected construction cost was at least US$5000/kW and the utilities had persuaded the government that 80 percent of the expected cost would have to be covered if finance was to be obtained and the scale of government guaranteed required for the same size of programme had increased 10-fold to US$30bn.[59] Similarly, the UK policy to build 15GW of new nuclear capacity by 2030 is not likely to be feasible unless guarantees worth more than £100bn are offered.[60]

2.3. Links to the Military Sector

Nuclear technology has its roots in military programmes. In many countries, a state-owned nuclear R&D organisation often with a dual civil/military remit and with significant influence in government has been an important force lobbying for the launch of reactor programmes. Such organisations included the U.S. AEC, the French CEA, and the UKAEA.

The first reactors were designed to produce plutonium for weapons via reprocessing of the spent fuel. Reprocessing was expected to be necessary for civil reactors because it was assumed that using uranium in power reactors would mean natural uranium reserves would be quickly depleted as only about 0.7 percent of naturally occurring uranium is 'fissile' (can support a chain reaction). This was expected to mean that reactors with fuel cycles that could utilise the non-fissile uranium would be needed. So-called fast reactors use plutonium as fuel obtained via a reprocessing plant.

Uranium enrichment is also a dual-purpose technology. Weapons can be produced using Highly Enriched Uranium (HEU), while most types of civil power reactors require the fissile naturally extracted uranium to be enriched from about 0.7 percent to about 3.5 percent.

---

59. S D Thomas (2010) 'The economics of nuclear power: an update' Heinrich Boell Stiftung, Brussels, pp 32-36.
60. S D Thomas (2016) 'The Hinkley Point decision: An analysis of the policy process' Energy Policy, 96, 421-431.

The most widely used civil reactor design, the Pressurised Water Reactor (PWR), was developed from reactors designed for military submarine propulsion units.

The skills required to handle nuclear materials are often common to civil and military uses so a strong civil programme provides a larger pool of skilled workers that the military sector can draw on.

These strong overlaps between military and civil nuclear technology are often seen as, at a minimum, using the civil nuclear sector to bolster the military sector and at worst, covertly using civil facilities for the military sector. For example, several of the newer weapon states such as India, Israel, and Pakistan acquired weapons plutonium from reprocessed fuel from reactors claimed to be civil research reactors. South Africa was able to make its nuclear weapon using an enrichment plant claimed to be for the civil sector and Iran's enrichment plants have been at the centre of concerns of weapons proliferation in Iran. Some analyses[61] claim that a key motivation of the current UK efforts to relaunch civil reactor ordering is to help retain the UK's capability in nuclear submarines. The key UK nuclear submarine company, Rolls Royce, said: 'One particular application for deployment of the talent developed through the UK SMR programme would be in the ongoing maintenance of the UK's independent nuclear deterrent. Currently, the UK Government is required to invest funding to sustain the skills and capability necessary for the maintenance of the Royal Navy's nuclear submarine programme.'[62]

---

61. P Johnstone & A Stirling 'Is Trident influencing UK energy policy' Sustainable Security. https://sustainablesecurity.org/2017/04/10/is-trident-influencing-uk-energy-policy-part-1/ and https://sustainablesecurity.org/2017/04/12/is-trident-influencing-uk-energy-policy-part-2/ (Accessed January 9, 2018.
62. Rolls Royce ' UK SMR: A national endeavour' Rolls Royce. https://www.rolls-royce.com/~/media/Files/R/Rolls-

3. China

3.1. Electricity Sector Structure

In many respects, China's nuclear sector and its electricity sector, in general, are more complicated than might be expected from an economy seen as highly centralised and planned, albeit the major actors are all nationally owned. New nuclear projects must be approved by the National Energy Administration (NEA) and the safety regulator, the National Nuclear Safety Administration (NNSA) but, especially in the past few years, few of the approved plants have gone ahead.

The electricity generation sector is dominated by five large regional companies with about 45 percent of the generation. Most of the rest of the capacity is owned by a large number of locally publicly-owned companies. The Big Five are Huaneng Group, Datang Corporation, Guodian Corporation, Huadian Corporation, and China Power Investment Corporation (CPIC). The vast majority of their capacity is coal-fired, with much of the coal coming from their own mines.

There are three reactor vendors, China National Nuclear Corporation (CNNC), China General Nuclear (CGN), and State Nuclear Power Technology Corporation (SNPTC), all of which also own and operate power plants, mostly nuclear. In 2017, CNNC announced a proposed merger with the China Nuclear Engineering Corporation (CNEC), the *de facto* monopoly nuclear construction company.[63] This merger would effectively reverse the split up in 1988 of the China Nuclear Industry group into CNNC, with the front-end and back-end fuel cycle and reactor design and supply, and

---

Royce/documents/customers/nuclear/a-national-endeavour.pdf (Accessed January 3, 2018).

63. Nuclear Intelligence Weekly 'CNNC-CNEC Merger Marks New Era for Chinese Industry' March 24, 2017.

CNEC, with an effective monopoly in civil and military construction. In January 2018, the merger appeared completed.[64]

While the Chinese home market has dominated the world market for nuclear power plants in the past decade, Table 1 shows its contribution to electricity supplies is tiny with only 2 percent of capacity nuclear and about 3.5 percent of power coming from nuclear power plants. Equally, the reactor vendors, which, along with Rosatom, are financially the most powerful in the world, are small compared to the Big Five utilities (see Table 2).

Table 1: China's generating capacity by source (2016)

Power source	Capacity (GW)
Thermal	1054
Water	332
Wind	149
Solar	77
Nuclear	34
Total	1646

Source: https://www.statista.com/statistics/302191/china-power-generation-installed-capacity-by-source/ (Accessed January 4, 2018)

The Chinese State Council has approved only CGN, CNNC, and SNPTC as nuclear developers with other companies, including the Big Five, allowed to take a minority stake in nuclear projects. SNPTC, the newest and weakest of the reactor vendors, has tried to overcome this problem by merging, in 2015, with the smallest of the Big Five, CPIC to form State Power Investment Corporation (SPI).

64. Nuclear Intelligence Weekly 'Weekly Roundup' February 2, 2018.

Table 2: Main utilities in China

Utilities	Installed capacity (GW)	Total assets $bn
The Big Five	663	
Huaneng	161	150
Guodian	135	122
Huadian	135	117
Datang	127	144
SPI (CPIC)	105	119
Nuclear vendors	42	
CGN	29	67
CNNC	13	68
SPI (SNPTC)	0	n/a
CNEC	0	9
Coal company		
Shenhua	78	144

Source: Nuclear Intelligence Weekly 'Reshaping the Nuclear Industry' April 14, 2017.

Note: SNPTC and CPIC merged to form SPI in 2015. None of the plants supplied by SNPTC was in service at the end 2017. The value of assets owned by the SNPTC part of SPI is not known.

Since then, there has been constant speculation about mergers in the power sector including between the Big Five and the other two reactor vendors, for example, CGN and Huaneng and CGN and Shenhua.[65] In June 2017, a proposed merger between Guodian and Shenhua was announced to form China Energy Investment Corp. In July 2017, a merger between Huaneng and SPI was announced, which, if completed, would reduce the Big Five to Four.

How far the decline in nuclear construction starts since 2010 is down to reluctance of the Big Five to be involved in nuclear projects is

65. Nuclear Intelligence Weekly 'Reshaping the Nuclear Industry' April 14, 2017.

hard to know. If the Big Five are reluctant, how far is this due to doubts about nuclear and how far is it due to a desire to be in control of their generating plants, not just minority shareholders?

3.2. Status of the Nuclear Programme

In 2018, China's nuclear capacity was 33.4GW (38 reactors), the fourth largest nuclear fleet in the world behind USA, France, and Japan.[66] It had a further 19 reactors (19.9GW) under construction, representing a third of the world's nuclear capacity under construction. Most of the operating reactors started construction in the period 2008-10 (see Table 3) when 25 of the 57 reactors were in service or under construction started construction. Since the beginning of 2011, perhaps before the Fukushima disaster, only 15 construction starts have taken place with ordering very uneven.

There may be a number of reasons behind the collapse in ordering from 2011 onwards apart from the reason normally quoted, the safety concerns the Fukushima disaster raised. This pause may, in part, reflect uncertainty about reactor choice. Most reactors use an old design licensed from the French company Areva but, from 2007 onwards, China has been attempting to move to more modern designs. It ordered six imported reactors in 2007-08 using two of the most modern designs available (four Westinghouse AP1000s and two Areva EPRs). A new Chinese vendor, State Nuclear Power Technology Corporation (SNPTC) was set up in 2007 to import and supply AP1000 technology while CGN was the partner for the EPR project. However, cost overruns and delays have occurred, with none of these yet in service and all six are five years late. The plan to move to one of these designs now appears in doubt. The other two

66. Data taken from the IAEA PRIS Database. https://pris.iaea.org/PRIS/home.aspx (Accessed May 17, 2019.

Chinese vendors, China National Nuclear Corporation (CNNC) and China General Nuclear (CGN), have been developing their own advanced designs since 2011 that they claim are their own intellectual property. Since 2013, there has been an attempt required by the Chinese government to merge these two designs. The result is known as Hualong One which is to be a unified design for export markets. Construction has started on eight reactors designated Hualong One. There have been difficulties in agreeing between CGN and CNNC on the elements of the merged design and the reactors under construction appear to comprise two somewhat different designs.

Since 2016, the Chinese government seems to have given priority to renewables over nuclear. In 2017, demand growth was back up to 6.5 percent and the contribution of renewables – hydro, wind, biomass, and solar – dwarfed that of nuclear. Solar output grew by 75 percent, wind, by 26 percent and biomass by 23 percent while nuclear output only grew by 16 percent. In response to the greater emphasis on renewables, the nuclear vendors, especially CGN, are diversifying into renewables, and in 2017, CGN added 15GW of new renewable capacity and it has plans for 13GW of off-shore wind.[67]

---

67. Nuclear Intelligence Weekly 'Nuclear Dwarfed by Renewables in 2017' January 26, 2018

Table 3: Reactor Construction starts in China

Period	No of reactors
1987-2007	17
2008	6
2009	9
2010	10
2011	0
2012	4
2013	3
2014	0
2015	6
2016	2
2017	0

Source:
https://www.iaea.org/PRIS/CountryStatistics/CountryDetails.aspx?current=CN
(Accessed January 2, 2018)

Table 4: Nuclear capacity in China (June 2016)

Province	Capacity (MW)
Guangdong	9378
Zhejiang	6564
Fujian	5445
Liaoning	3356
Jiangsu	2120
Guangxi	1086
Hainan	650

Source: https://www.statista.com/statistics/302191/china-power-generation-installed-capacity-by-source/ (Accessed January 4, 2018)

The problems with the imported AP1000s and EPRs have continued. The two EPRs were originally scheduled for completion in 2013/14 but by 2017, this had slipped to 2018. However, an equipment problem made public in December 2017 will lead to further delays, not quantified at the time of writing.[68] The four AP1000s were claimed not to be affected by the bankruptcy of Westinghouse but none are yet in service and all are 4-5 years late.

Other factors limiting orders include the *de facto* ban on building reactors on inland sites, the reduction in electricity demand growth, and the strain on human resources that starting construction on 25 reactors in the three years from 2008-10 imposed. Nuclear reactors are installed in only seven of China's 34 provinces with more than three-quarters of the capacity installed in Guangdong, Zhejiang, and Fujian (see Table 4). There have been frequent forecasts since 2011 that the pace of construction would pick up soon, but these have not been fulfilled. In February 2017, the National Energy Administration announced plans to approve and start construction on eight new reactors. In the event, no construction starts were made in 2017.[69]

The problems with the imported AP1000s and EPRs have continued. The two EPRs were originally scheduled for completion in 2013/14 but by 2017, this had slipped to 2018. However, an equipment problem made public in December 2017 will lead to further delays, not quantified at the time of writing.[70] The completion of the four AP1000s was claimed not to be affected by the bankruptcy of Westinghouse but none are yet in service and all are five years late.

68. Nuclear Intelligence Weekly 'Taishan EPR Set for Further Delays' December 15, 2017.
69. Nuclear Intelligence Weekly 'China Plans Eight Reactor Construction Starts in 2017' February 17, 2017.
70. Nuclear Intelligence Weekly 'Taishan EPR Set for Further Delays' December 15, 2017.

3.3. Electricity Price Setting

The electricity sector has come under great strain since 2013 due to a combination of a sharp reduction in electricity demand growth leading to major over-capacity and the need to deal urgently with the air quality crisis in the cities by cleaning up and reducing coal-fired generation. Overcapacity is a particular problem in Liaoning, Fujian, and Hainan provinces, all of which have nuclear capacity. This strain and the reduction in profits of the Big Five is a major factor behind the merger activity in the electricity sector. The nuclear sector is a very small player in the electricity sector. The nuclear companies are small compared to the Big Five and the coal company, Shenhua. Even the most optimistic forecasts of nuclear expansion over the next decade or two made by the Chinese government would still leave nuclear supplying only a small percentage of China's power.

Nevertheless, the strains on the electricity sector have had a serious impact on the viability of the nuclear sector. The government sets the rate utilities must pay for the 'guaranteed output' of the reactors, in 2017 43 Chinese cents/kWh ($6.31c/kWh). In the past, this guaranteed output was set as the equivalent of about 7000 hours (80 percent of the year) of operating time so nearly all the output was sold at the set rate.[71] However, from 2015, provincial authorities have been allowed to set up regional trading markets and have the ability to set their own rules for nuclear generation including the guaranteed output, but not the price paid. First-of-a-kind reactors, such as EPR, AP1000, and Hualong One, will be able to sell all their output at the guaranteed price. The Guangxi province set the guaranteed feed-in time at only 4800 hours, leaving the utilities to bid the rest of their power into the market. Provincial authorities are

71. Nuclear Intelligence Weekly 'Reactor Operators Facing Competitive Markets' June 16, 2017, p 4.

also exerting pressure for the guaranteed price to be reduced.[72] The market price, generally set by the marginal price of coal-fired generation, is substantially below the guaranteed price. As a result, load-following for nuclear plants has had to be introduced and, in one case, the start-up of a completed reactor was delayed because of low demand for the output. The utilisation rate of reactors is therefore falling significantly.[73]

3.4. Availability of Low-Cost Finance

No details are known on how the Chinese nuclear programme is financed. Given that the companies involved are state-controlled and given the high level of currency reserves in China, it seems likely that finance was readily available and the cost low.

3.5. The Military Sector

In China, the exploitation of nuclear technology was originally through the Ministry of Nuclear Industry which was reorganised and became CNNC and CNEC in 1988 and the military activities are now based almost entirely in CNNC and CNEC, which were in the process of re-merging in early 2018.[74] The weapons capability substantially pre-dates the civil nuclear programme by about 20 years to the mid-60s. In the past, CNNC's website has prominently claimed it was the organisation that built China's atom bomb, hydrogen bomb, and nuclear submarines. It is now more discreet in

72. Nuclear Intelligence Weekly 'Power Market Liberalization — More Challenges to Nuclear?' June 24, 2016.
73. Nuclear Intelligence Weekly 'A Reality Check for Nuclear Power — Lower Demand' June 10, 2016 and Nuclear Intelligence Weekly 'China's 13th Five-Year Plan — Mission Impossible?' November 4, 2016.
74. Nuclear Intelligence Weekly 'Weekly Roundup' February 2, 2018, p 1.

its claims and only says it is 'a leading element of national strategic nuclear forces.'[75] It is difficult to know how far public opposition that halted the development of a reprocessing plant in Jiangsu province in 2016 was due to the military connection or whether it was simply down to safety concerns.[76]

3.6. Assessment

China's electricity industry appears to be an uncomfortable compromise between a centrally planned system and a market system without the control of a centrally planned system or the competitive pressures a full market system would bring. Like the USA and many European utilities 40-50 years ago, Chinese utilities failed to spot the strong downturn in demand growth as demand began to saturate and the economy started to move away from energy intensive industries. As a result, China now faces increasing generation over-capacity leading to little need for new capacity to meet growth as the air quality crisis, and the need to reduce greenhouse gas emissions leads to the need to build new low-carbon capacity to replace coal generation, much of which is relatively new.

In some respects, given the minimal contribution of nuclear to China's existing and almost certainly its future electricity supplies, nuclear power is a sideshow. However, from the perspective of the global nuclear industry, the future of the Chinese nuclear industry is key. In short, the world nuclear industry needs China but China does not need nuclear power. The nuclear military sector in China seems

75. CNNC 'Company Profile' CNNC, Beijing. http://en.cnnc.com.cn/2016-02/01/c_49164.htm (Accessed January 5, 2018).
76. International New York Times 'Chinese city withdraws nuclear plan after protests' August 12, 2016.

well enough established that the support of a civil programme is not essential.

There appear to be four major barriers to new nuclear orders in China: the *de facto* ban on building nuclear reactors at inland sites; generation overcapacity particularly in the regions where nuclear is feasible; problems of cost and buildability with the modern designs being developed; and the apparent lack of interest amongst the Big Five generators in nuclear power.

Central government sets the selling price for the guaranteed output of the nuclear plant. This price, US6c/kWh, appears low by standards of other countries particularly for the latest designs: AP1000, EPR, and Hualong One. For example, the UK is expecting to pay double this amount for the power from the proposed Hinkley Point C EPRs. It is not clear yet whether Hualong One will prove cheaper and more buildable than the imported designs. There are no reports of major construction problems yet, but at this stage in the construction of the AP1000s and the EPRs in China, there was no acknowledgement that things were going wrong. There is also a question mark about how much output of the old designs will be paid this price and how much will have to accept the much lower market price set in the new regional markets. The level of the guaranteed output is now in the hands of the provincial authorities rather than central government and the provincial authorities may have an incentive to reduce the guaranteed hours to reduce the prices consumers must pay.

4. Taiwan

4.1. Electricity Sector Structure

The electricity sector in Taiwan is dominated by Taipower, which is a fully integrated state-owned utility established in 1946.[77] Nevertheless, independent power producers (IPPs) do contribute about a quarter of Taiwan's power supplies. In 2016, three-quarters of Taiwan's capacity was fossil fuel, about 10 percent nuclear and the rest renewables. Installed capacity has doubled in the past 20 years but nuclear capacity has remained static so the contribution of nuclear to total capacity has halved. Plant closures and shutdowns from 2014 onwards have meant that the contribution of nuclear to generation has fallen from 26.5 percent in 1995 to 12 percent in 2016.[78]

There are plans to install 20GW of new generating capacity by 2026, two-thirds of which would be renewables, mostly owned by IPPs. However, a quarter of the new capacity is expected to be coal-fired.

There are plans to unbundle Taipower, introduce a competitive wholesale electricity market, give consumers choice of supplier, and introduce an Electricity Regulatory Authority (ERA) but this is all in the future.

---

77. Chuan-Neng Lin 'Developments in Taiwans'e electricity market' Bureau of Energy, Taipei. https://www.emcsg.com/f1671,123955/3_-_Dr_Chuan-Neng_Lin_Bureau_of_Energy_Taiwan.pdf (Accessed January 10, 2018).
78. World Nuclear Association 'Nuclear Power in Taiwan' World Nuclear Association. http://www.world-nuclear.org/information-library/country-profiles/others/nuclear-power-in-taiwan.aspx (Accessed May 17, 2019).

4.2. Status of the Nuclear Programme

In 2018, the IAEA reported that six reactors (5.1GW) were in service in Taiwan with two (2.6GW) more under construction (see Table 5).[79] However, work on the two under construction was suspended in December 2014, 15 years after construction started, and is unlikely to recommence. In January 2016, a new government was elected with a promise to phase out nuclear power by 2025, retiring the plants after 40 years of operation.

Table 5: Reactor Construction starts in Taiwan

Year	**Reactors**	**Commercial operation**
1972	Chinshan 1	1978
1973	Chinshan 2	1979
1975	Kuosheng 1	1981
1976	Kuosheng 2	1983
1978	Maanshan 1	1984
1979	Maanshan 2	1985
1999	Lungmen 1 & 2	-, -

Source:
https://www.iaea.org/PRIS/CountryStatistics/CountryDetails.aspx?current=TW
(Accessed January 2, 2018)

The completed reactors all started construction in the 1970s, with a gap of nearly 20 years before the two uncompleted reactors started construction. Four of the six completed reactors (two at both of the Chinshan and Kuosheng sites) and the two uncompleted reactors (at Lungmen) are Boiling Water Reactors (BWRs) supplied by General

79. http://www.world-nuclear.org/information-library/country-profiles/others/nuclear-power-in-taiwan.aspx (Accessed January 10, 2018).

Electric (GE), while the other two (at Maanshan) are Pressurised Water Reactors (PWRs) supplied by Westinghouse.

The Lungmen reactors have been delayed almost since the start of construction.[80] The original completion date when the contract was awarded was 2004, but in 2001 the project was delayed by the government for three months and the expected completion date was put back to 2006. In 2004, the schedule was delayed by a further 18-24 months and by 2005, completion was not expected until 2010. The reasons cited were escalating prices and government procurement rules. Delays in the government providing funding continued in 2006. By then, a nuclear phase-out in Taiwan was on the agenda. In 2009, completion was not expected until 2012. The problems were not confined to funding and in 2011, the Taiwan nuclear safety authority was considering suspending construction due to construction difficulties and project quality.[81] In 2014, the government decided to suspend work at the site.[82] Unit 1 was then said to be largely complete and unit 2, 90 percent complete. It was reported that Taiwan had spent US$9.9bn on these plants.[83] In December 2015, GE began arbitration proceedings with Airpower to recover its costs.[84] Following the election of a government

80. Nucleonics Week 'Finish of Taiwan's Lungmen ABWRs slips again, this time to 2009/2010' May 5, 2005.
81. Nucleonics Week 'Taiwan's regulator undecided on suspending Lungmen construction' August 11, 2011.
82. Nucleonics Week 'Taiwan government suspends Lungmen work in face of protests' May 1, 2014.
83. Nuclear Engineering International 'World survey - Nuclear new-build - New-build now' June 2014.
84. World Nuclear News 'GE seeks arbitration over Lungmen payments' World Nuclear News, London http://www.world-nuclear-news.org/C-GE-seeks-arbitration-over-Lungmen-payments-1412154.html (Accessed January 12, 2018).

pledged to phase out nuclear power in 2016, there appears little chance the reactors will now be completed.[85]

The oldest reactor, Chinshan 1, has not operated since 2014, initially due to an equipment failure but subsequently due to lack of spent fuel storage capacity, and this closed the other unit in 2016. Neither reactor, scheduled for closure in December 2018 and July 2019 respectively, is likely to operate again. The second unit at the second plant, Kuosheng, due to be closed in March 2024, was also closed in summer 2016 due to a fire in the turbine and may also not restart. The first Kuosheng unit is due to close in December 2022 and the Maanshan reactors are due to close in July 2024 and May 2025.[86]

4.3. Electricity Price Setting

There appears to be no independent and rigorous regulation of electricity prices.

4.4. Availability of Low-Cost Finance

As a state-owned company, Taiwan Power would have had access to public funds for its nuclear programme and given its *de facto* monopoly status, it would have had access to low-cost finance.

---

85. Nucleonics Week 'Taiwan president-elect reaffirms goal of 'nuclear-free' Taiwan by 2025' March 17, 2016.
86. Nuclear Intelligence Weekly 'Government Boosts Pace Toward 'Nuclear-Free Homeland'' September 9, 2016.

4.5. The Military Sector

In the 1970s and 1980s, there were serious concerns about Taiwan's intentions to develop nuclear weapons through the Institute of Nuclear Energy Research (INER), which remains in operation, albeit with little suggestion of any military agenda.[87] Taiwan's situation is complicated by the fact that since 1971, it has been categorised by the UN and thus the IAEA as a province of China, not a sovereign state and its signature in 1968 to the Nuclear Nonproliferation Treaty is no longer valid. Whether statements from officials suggesting development of nuclear weapons are any more than rhetoric is hard to know.

4.6. Assessment

Taiwan's nuclear programme dates from the period before democratic rule was introduced and a centralised authoritarian government that owned and controlled the electricity sector was able to force through a nuclear programme. Public opposition has grown against nuclear power, and the plans in 2018 were to phase out nuclear power by 2025. Experience in other countries that have adopted nuclear programmes suggests that while this timetable might slip, the chances of a policy reversal and new orders being placed are minimal.

---

87. Nuclear Threat Initiative 'Taiwan and proliferation' NTI, Washington. http://www.nti.org/analysis/articles/taiwan-and-nonproliferation/ (Accessed January 10, 2018).

5. *Saudi Arabia*

5.1. Electricity Sector Structure

The dominant company in the electricity sector is the Saudi Electricity Company. It owns about 74GW of capacity, about 60 percent of which is gas-fired and the rest oil-fired.[88] It is a *de facto* monopoly, fully integrated generation/transmission/distribution/ retail company. It was established in 2000 from the merger of all Saudi electricity companies. The Saudi government owns about 80 percent of the shares, mostly directly, with some owned by Saudi Aramco. There are plans to functionally unbundle the company into competing generation companies, a transmission company, and distribution/retail companies. There are a small number of independent power producers, allowed from 2007 onwards.

Electricity demand appears to be growing rapidly, and power generated in 2016 was 7 percent more than in 2015. The government anticipates rapid growth will continue and in 2016, the King Abdullah City for Atomic and Renewable Energy (KA-CARE) was calling for an additional 41GW of solar power, 17.6GW of nuclear, and 9GW of wind power by 2032, although the nuclear target had already been downgraded by the government.[89]

---

88. Saudi Electric Company 'Information about company' SEC, Riyadh. https://www.se.com.sa/en-us/Pages/aboutus.aspx and Energy Information Administration 'Saudi Arabia', EIA, Washington. https://www.eia.gov/beta/international/analysis.cfm?iso=SAU (Accessed January 9, 2018).
89. KA-CARE 'Saudi Arabia's Sustainable Energy Program in a Nutshell' KA-CARE' Riyadh. https://ksa-climate.com/wp-content/uploads/2018/11/Saudi-Arabias-sustainable-energy-program-in-a-Nutshell.pdf (Accessed May 17, 2019).

5.2. Status of the Nuclear Programme

The attempt to launch a nuclear power programme dates from August 2009, when the Saudi government announced that it was considering a nuclear power programme.[90] At that time, KA-CARE was set up to lead the nuclear programme. In April 2010, a royal decree was issued saying: 'The development of atomic energy is essential to meet the Kingdom's growing requirements for energy to generate electricity, produce desalinated water and reduce reliance on depleting hydrocarbon resources.'[91]

The potential size of the Saudi reactor market and its ability to provide its own finance makes Saudi Arabia a very attractive market for all the reactor vendors, including small reactors for desalination purposes. A 2013 target of 17GW of nuclear capacity by 2030 had slipped to 2040 by 2015 and by 2017, it was not clear whether there was still any target. The first step would be the ordering of two reactors (2.2-3.2GW) at a site yet to be determined. In January 2018, two sites, Umm Huwayd and Khor Duweihin, were reported to have been shortlisted.[92]

Vendors who have expressed an interest in the market include GE-Hitachi, Toshiba/Westinghouse, Areva, and Rosatom. In October 2017, KA-CARE submitted a request for information (RFI). Several vendors have expressed an interest in the market but those responding to the RFI were reported to be China National Nuclear

---

90. World Nuclear Association 'Nuclear power in Saudi Arabia' WNA, London. http://www.world-nuclear.org/information-library/country-profiles/countries-o-s/saudi-arabia.aspx (Accessed January 10, 2018).
91. World Nuclear Association 'Nuclear power in Saudi Arabia' WNA, London. http://www.world-nuclear.org/information-library/country-profiles/countries-o-s/saudi-arabia.aspx (Accessed December 14, 2017).
92. Middle East Business Intelligence 'Saudi Arabia shortlists two sites for first nuclear power project' January 22, 2018. https://www.meed.com/exclusive-saudi-arabia-shortlists-two-sites-first-nuclear-power-project/ (Accessed February 5, 2018).

Corporation, Korea Electric Power Co (KEPCO), EDF/Framatome, Rosatom, and Westinghouse.[93] The financial collapse of Areva and Westinghouse mean it may be problematic for them to place bids. By end 2017, Saudi Arabia was claiming it would launch reactor tendering process within months with a view to selecting the vendor by end 2018, signing a contract in 2019 and commissioning the first unit in 2027.[94] This timetable is very ambitious.

Suppliers mentioned for the small reactor market include INVEP (Argentina) with its 27MWe CAREM reactor (under construction in Argentina since 2014); China Nuclear Engineering Corporation with its high temperature reactor, HTR-PM, based on the reactor under construction in China; and KAERI (Korea) with its SMART reactor (up to 100MWe), to be jointly developed with KA-CARE. KA-CARE is reported to own a percentage of the Intellectual Property for the SMART design.[95]

5.3. Electricity Price Setting

A regulatory body, the Electricity and Cogeneration Regulatory Authority (ECRA), was set up in 2001.[96] Its mission is to 'insure [sic] that supplies of electricity and desalinated water products provided to consumers in the Kingdom are adequate, reliable, of high quality and reasonably priced'. Its most recently published

---

93. Nuclear Intelligence Weekly 'Ka-Care Hopes to Choose from Five Bids by Year's End' January 19, 2018.
94. Nuclear Intelligence Weekly 'Reactor Vendors Poised to Bid on Saudi RFI' December 15, 2017, p 4.
95. Middle East Business Intelligence 'Saudi Arabia shortlists two sites for first nuclear power project' January 22, 2018. https://www.meed.com/exclusive-saudi-arabia-shortlists-two-sites-first-nuclear-power-project/ (Accessed February 5, 2018).
96. Electricity & Cogeneration Regulatory Authority 'About ECRA' ECRA, Riyadh. http://www.ecra.gov.sa/en-us/Pages/default.aspx (Accessed January 9, 2018).

annual report was for 2014.[97] It states that its responsibilities include:

> 'Assessment of tariffs charged for supply of the electricity and water desalination industry services, periodic review of these tariffs and when modification is needed, modifying whatever tariffs fall within its purview while for other tariffs submitting proposals to the government.'

It is difficult to determine how rigorous and independent these reviews are and how far they determine electricity tariffs.

5.4. Availability of Low-Cost Finance

Given that SEC is state-controlled and given the vast income Saudi Arabia has from oil and gas, it seems likely that whoever was charged with owning and operating any Saudi reactors would have access to cheap capital.

5.5. The Military Sector

Saudi Arabia's avowed intention to build up a full fuel cycle, including enrichment and reprocessing, to match its reactor programme has led to serious concerns about the proliferation risk this raises. In particular, there are concerns that the U.S. Trump government wants to come to an agreement with Saudi Arabia that

97. Electricity & Cogeneration Regulatory Authority 'About ECRA' ECRA, Riyadh. http://www.ecra.gov.sa/en-us/MediaCenter/doclib2/Pages/SubCategoryList.aspx?categoryID=4 (Accessed January 9, 2018).

does not require it to fulfil the 'Gold Standard' of a promise not to reprocess fuel and extract plutonium nor to enrich uranium.[98]

5.6. Assessment

Saudi Arabia has yet to commit to a nuclear programme and, given the wealth of the country and its ability to provide cheap finance, winning orders in Saudi Arabia would represent a major coup for a successful vendor. Despite this, perhaps only the Chinese vendors, who are only mentioned in the context of supplying small reactors using their unproven high-temperature gas-cooled reactor technology, are in a position to sell reactors at loss-leader prices and even they would not be able to sell at a loss for more than one or two reactors. So, unless there is a strong covert agenda to develop a military nuclear capability under the cover of a civil programme, there must be serious doubts about whether Saudi Arabia will proceed with a nuclear programme when it sees the scale of costs involved. Intuitively, Saudi Arabia would appear an ideal location to develop solar power with the demand profile (notably for air conditioning) and solar supply availability appearing to match well.

98. The Hill 'Reactors for Saudi Arabia are bad business and dangerous diplomacy' January 7m 2018. http://thehill.com/opinion/national-security/367838-reactors-for-saudi-arabia-are-bad-business-and-dangerous-diplomacy (Accessed January 9, 2018).

6. *Iran*

6.1. Electricity Sector Structure

By 2017, the installed electricity capacity in Iran had reached 77GW.[99] Until 2000, the Iran electricity sector was dominated by the state-owned vertically integrated company, TAVANIR.[100] Since then there has been a process of privatisation and introduction of competition so that by 2015, nearly half of Iran's power generation came from privately owned companies. However, TAVANIR remains the main player and its website states it 'manages generation, transmission and distribution of power'.[101]

About two-thirds of Iran's generating capacity uses natural gas and about a quarter uses oil, with nuclear, coal, hydro, and renewables making up the remaining 10 percent. Demand has grown rapidly in recent years, by about 7 percent per annum leading, at times, to shortages. However, in 2014, the government increased prices by 25 percent and again by 20 percent in 2015 to reduce subsidies and to dampen growth.[102]

There are 16 Regional Electricity Companies (RECs), all wholly owned by TAVANIR, which have a management role in their region. There are 39 Distribution Companies (DCs) that are responsible for distribution and retail in their region. 40 percent of the shares in each

99. Iran Daily 'Iran's power generation capacity reaches about 77,000MW' June 10, 2017. http://www.iran-daily.com/News/194482.html (Accessed January 15, 2018).
100. G.RezaYousefi, Sajjad Makhdoomi Kaviri, Mohammad Amin Latify, Iman Rahmati (2017) 'Electricity industry restructuring in Iran' Energy Policy, volume 108, pp212-226.
101. Tavanir 'About Tavanir :: Historical background of Electric Power Industry in Iran' Tavanir, Tehran.
http://www.tavanir.org.ir/en/1.php (Accessed January 10, 2018).
102. Data is taken from: Energy Information Administration 'Iran' EIA, Washington. https://www.eia.gov/beta/international/analysis.php?iso=IRN (Accessed May 17, 2019).

of these companies is owned by TAVANIR with the balance owned by a public/private company established in 2004, SABA. There are 39 Power Generation Management Companies that operate the power plants in their region. Again, 40 percent of the shares are owned by TAVANIR, 40 percent by the Ministry of Energy, and 20 percent by others. There are many generation companies, some owned by TAVANIR, some by private investors. There is a pool-based day-ahead wholesale power market, the Iranian Electricity Market (IEM), established in 2003. It is not clear how far this market places genuine competitive pressure on the companies given that it can be effectively bypassed with bilateral contracts.

A regulatory body, the Iran Electricity Market Regulatory Board (IEMRB), was established in 2004, but its role appears to be setting and monitoring rules rather than regulating prices.

6.2. Status of the Nuclear Programme

Iran had ambitious nuclear plans until 1979 under the Shah.[103] Two 1200MW PWRs were ordered from the German company Kraftwerk Union (later known as Siemens) in 1976 to be built at the Bushehr site. Work started on the plant then and was claimed to be 77 percent complete when work was abandoned in 1979 after the overthrow of the Shah.[104] Two PWRs were ordered from the French company, Framatome, for the Karun River site (also known as Darkhovin and Esteghlal) in 1977 but were abandoned a year later.[105] In 1991, a Swiss arbitration court awarded Framatome

103. World Nuclear Association 'Nuclear power in Iran' WNA, London. http://www.world-nuclear.org/information-library/country-profiles/countries-g-n/iran.aspx (Accessed January 10, 2018).
104. Washington Post 'Iran Set to Scrap $34 Billion Worth of Civilian Projects' May 30, 1979.
105. Washington Post 'Iran Cuts Back Nuclear Plans To Save Money' October 23, 1978.

US$700m in compensation for cancellation of the contract.[106] Four more reactors were ordered from Kraftwerk Union but cancelled before work had started. Around this time, Iran took a stake in the Eurodif uranium enrichment plant built in France.

Plans to complete the Bushehr plants were discussed with Siemens in the 1980s but in 1987, the facility was bombed during the Iran/Iraq war.[107] Despite this, discussions continued with Siemens over the completion of the reactors. However, in 1995, Iran reached an agreement with Russia to complete one of the reactors.[108] Construction restarted in 1996 but commercial operation was not until 2013. The single reactor appears to be a V-320 Rosatom reactor (the Russian standard model from the 1980s) housed in the existing Siemens containment (Khlopkov & Lutkova, 2010). Operation of the plant has been problematic. It took 28 months from first power to commercial operation (typically this takes about three months) and in its three full years of commercial operation to end 2016, its capacity factor was only 53 percent compared to the world average of about 80 percent.[109]

Discussions on further orders did not take place until the Bushehr unit was in commercial operation. In 2014, Iran agreed to a deal with Rosatom to build two further reactors at the Bushehr site with an option for six further reactors, two more at Bushehr and four at an undetermined site. There was optimism that construction would start soon but in November 2017, the pouring of the first structural concrete, the normal point at which construction is marked as being

106. Agence France Presse 'Iran told to pay France in 700 million in damages' October 2, 1991.
107. United Press International 'Iraq bombs unfinished Iranian nuclear power plant' November 17, 1987.
108. Xinhua News Agency 'Iran, Russia sign nuclear power contract' January 8, 1995.
109. Annual capacity factor is the power produced in a year as a percentage of the power that would have been produced had the plant operated at full power uninterrupted.

underway, was not expected before 2019 with operation in 2026 and the second unit following two years later.[110]

It also signalled its interest in building an indigenous PWR design of 360MW at the Darkhovin site and buying the ACP100 SMR from China (CNNC) for installation at the Makran site.[111] There is no schedule for either of these projects and, if they go ahead, start of construction is several years away.[112]

6.3. Electricity Price Setting

It is clear that it is the government, not IEMRB, that sets prices and, given the dominance of the state-owned TAVANIR, this is not surprising.

6.4. Availability of Low-Cost Finance

While Iran has huge reserves of oil and gas, international sanctions have seriously reduced its sales of oil and gas, and the availability of finance is heavily dependent on the world oil price. With the relatively low oil prices of 2017/18 and sanctions still in force, it seems likely that an Iranian nuclear programme would only be possible with foreign finance.

---

110. Nuclear Intelligence Weekly 'Briefs' November 3, 2017.
111. Nuclear Intelligence Weekly 'Iran Looks to China as Well as Russia for Reactors' April 17, 2015.
112. Nuclear Intelligence Weekly 'Potential and Existing Global Nuclear Newbuild Projects (Generation III+ or Earlier)' September 8, 2017 and Nuclear Intelligence Weekly 'Limited Newbuild Ambitions Beyond Bushehr' October 14, 2016.

6.5. The Military Sector

There has been long-running concern about Iran's uranium enrichment and reprocessing plants, even though Iran claims it is in compliance with the Nonproliferation Treaty. The Tehran Nuclear Research Centre (TNRC) has always been at the heart of Iran's civil and military nuclear capability including the enrichment and reprocessing facilities.

6.6. Assessment

Iran has a long history, going back more than four decades, of trying to develop nuclear power. All it has to show for this is one rather problematic operating reactor. Proliferation concerns would make it difficult for US, European, and Japanese vendors to supply plants there. Iran says it has its own reactor technology but over the last decade, claims that construction would start soon on an Iranian-designed reactor have not been realised.

Iran's experience with the Russian nuclear industry, which took nearly 20 years to complete the part-built reactor at Bushehr, seems poor, so it is surprising that Iran has turned to Russia for new orders. China has signalled its interest in the Iranian market, but Iran appears only interested in its unproven Small Modular Reactor design, ACP100, rather than Hualong One.

Construction start with Russian reactors is still reported to be at least two years off and the deal may yet collapse. Even if construction does start, the pace of ordering for follow-up orders is expected to be slow. Iran's apparent desire for technologies with dual civil/military applications may be a factor that will force an uneconomic nuclear programme through.

Meeting China's Future Energy Needs: Is Increasing Nuclear Power the Best Way to go?

David von Hippel

Executive Summary

After three decades of rapid economic growth, with attendant increases in energy use and in the environmental consequences of same, as of 2018 China arguably sits at a point of decision, inflection, or possibly both in the evolution of its electricity generation system, its nuclear power future, and possibly the future of its energy sector as a whole. As a source of electric power that is essentially free of the air pollutant and greenhouse gas emissions that plague the coal-fired generation that produces up about 70 percent of China's electricity, nuclear power has been touted as a solution to China's environmental challenges. Scenarios of nuclear power development in China, including those presented in this paper, span the range from modest growth to 80-100 GWe (gigawatts of electricity generation capacity) by 2050 (from about 34 GWe today) to growth to 300 to 400 or more GWe of nuclear power by 2050. China's nominal plans for its nuclear sector include the use of spent fuel reprocessing to extract plutonium (Pu), the use of mixed-oxide (MOx—a mixture of plutonium and uranium isotopes) nuclear fuel in light water reactors, and ultimately, the development and use of fast-breeder reactors that produce and use plutonium fuel. These technologies potentially pose risks for the international nonproliferation regime by creating large stocks and flows of plutonium that, if diverted, could be used in nuclear weapons or "dirty bombs." This paper explores the potential for China to meet its development needs and environmental goals through three alternative paths of nuclear power development through 2050 and explores the relative physical, economic, environmental, and socio-political consequences of the paths.

A comparison of alternative paths for the Chinese nuclear and electricity sectors—namely a "BAU" (Business As Usual) or reference path reflecting recent trends and plans, a "MAX" path with fairly aggressive expansion of nuclear generation capacity, and a "MIN" path in which nuclear power capacity expands more slowly, and construction largely stops after 2040, with energy efficiency and renewable energy sources becoming a major focus—suggests that it will be possible to meet China's economic development, GHG and air pollutant emissions reduction, and other goals without an extended or massive build-out of LWR capacity, and without expansions of uranium enrichment or reprocessing capacity beyond projects now underway.

In particular, the "Minimum Nuclear" (MIN) path means that China will likely not become a major exporter of nuclear power technologies but will continue along its current trend of being perhaps a dominant provider of renewable power systems. In addition, the MIN path nuclear sector costs are much lower (both in aggregate and per unit output), than in the MAX and BAU paths. Although nuclear costs are only a small part of overall cost of providing energy services to the Chinese economy, past experience in China and elsewhere, as well as other studies of Chinese energy futures, have indicated that emphases on energy efficiency and renewable energy offer China the ability to effectively address environmental concerns without significant (if any) additional costs, relative to a reference path. In addition, a path with less nuclear power, and less nuclear fuel supply and enrichment ("front-end") and nuclear spent fuel handling, transport, and reprocessing ("back-end") facilities is arguably easier to deploy in social and political terms, particularly if expectations for a stronger voice in how China's future unfolds continue to grow among the Chinese citizenry (although that is not a given as of this writing). The MIN path provides significant benefits over MAX/BAU paths in terms of Pu production and stocks, and thus offers a significantly lower risk of nuclear weapons proliferation than the other two paths.

In order for the MIN path (or a similar energy trajectory) to become reality in China, national policy support for energy efficiency and renewable energy will need to take precedence over policy support for nuclear power and technological and cost trends in energy efficiency and renewable energy will need to continue or accelerate. Trends in recent years have pointed toward the enhanced practicality of a low-nuclear path for the Chinese energy sector. These trends include a slow-down in reactor construction, and a re-thinking of nuclear safety regulations in the post-Fukushima era, ongoing structural change in the Chinese economy away from heavy industry, with attendant much-reduced growth in electricity needs, and exceedance of even the State's own ambitious targets for renewable energy.

The nations of the international community can help to influence a Chinese transition to a low-nuclear future through the acknowledgment of the benefits of a MIN path (or similar) for China by international political and trading partners, through international policies encouraging low-nuclear paths, and through embracing energy paths of their own that de-emphasize nuclear power, enrichment and reprocessing, and encourage nuclear sector safety and transparency.

Acknowledgements

The author acknowledges and thanks the Nonproliferation Policy Education Center for its support for the development of this paper, and in particular thanks Henry Sokolski for his ideas in framing the paper and his review comments and Stephen Thomas for his review of and additions to the paper.

Many of the energy security analytical frameworks and methods used to develop and evaluate the energy paths described in this report were assembled as a part of the Asian Energy Security and related projects carried out by the Nautilus Institute for Security and Sustainability over the last two decades, with funding from the MacArthur Foundation and other donors. The author is grateful to have had the opportunity to work extensively with colleagues from Nautilus and with many colleagues from East Asia in developing and applying these energy security approaches.

1. Introduction: China's Energy and Environmental Challenges

1.1. Projections of Economic Development and Energy Use in China, and Their Environmental Price

China's emergence as a global economic powerhouse from what was a largely rural and agrarian society even as of the 1980s has been a defining characteristic of the past few decades. The rate of growth in gross domestic product (GDP) in China exceeded 7 percent per annum in each year between 1991 and 2015, dipping only slightly below 7 percent in 2015 and 2016.[113] The result of this rapid growth has been that China's economy is more than 11 times larger today than it was in 1990, and nearly 30 times larger than it was in 1980. An increase in energy production and use, and especially electricity generation, has fueled the growth in China's economy. Primary

113. World Bank (2017), "GDP Growth, Annual (%)", graph and data available at http://data.worldbank.org/indicator/NY.GDP.MKTP.KD.ZG?locations=CN.

energy use, a measure of all of the fuels that go into providing energy for an economy, has increased by nearly a factor of five since 1990, as has oil and oil products consumption, and electricity generation has grown nearly 10-fold in the same period. China edged out the United States as the world's largest producer of coal in the mid-1980s. In 1990, coal production in China was over a billion tonnes (metric tons) per year, and rose to nearly 4 billion tonnes by 2013 before declining somewhat in recent years.

Although it has significant energy resources of its own, China has been obliged to turn to imports to help fuel its massive economic growth. China became a net importer of oil and oil products in the early 1990s, and a net importer of gas, mostly as liquefied natural gas (LNG), around 2006. By 2016, China was the world's third-largest importer of LNG, after Japan and the ROK. China's oil imports broke a monthly record early in 2017, with its imports exceeding those of the U.S. to be the world's highest.[114] China sources its oil imports from a diverse array of nations; by 2014, 14 countries each supplied at least 2 percent of China's total oil imports, with Saudi Arabia its leading supplier at 16 percent.[115] China's energy imports dependency continues to increase as consumption continues to rise while production of coal and oil, in particular, are relatively static. China's energy imports dependency is not yet, however, at the 90-plus percent level found in the ROK and Japan.

The large and rapid increase in China's energy use has been accompanied by a variety of environmental challenges. In 2006, China passed the United States to become the largest emitter of

---

114. See, for example, Gordon Kristopher (2017), "China's Crude Oil Imports Hit a New Record
Market Realist, April 19, 2017, available as
http://marketrealist.com/2017/04/chinas-crude-oil-imports-hit-new-record/.
115. United States Department of Energy, Energy Information Administration (2015), *China*, last updated May 14, 2015, and available as
https://energy.gov/sites/prod/files/2016/04/f30/China_International_Analysis_US.pdf.

greenhouse gases (GHGs) among nations,[116] and by 2014 emitted 30 percent of global anthropogenic GHG emissions, although its per capita emissions remain lower than those in the U.S. and many other industrialized nations.[117] In addition, increased energy use in China has led to emissions of local and regional air pollutants high enough to be a significant danger to health during much of the year in many Chinese cities, including Beijing,[118] and has contributed to water pollution, soil degradation, and numerous other environmental problems.

Although China has markedly improved energy use efficiency in recent years, particularly as measured in energy use per unit of GDP, shifting away from coal use to reduce global, regional, and local environmental problems has continued to be a priority in Chinese policy. This shift is to be in part accomplished by moving some of the most polluting industries out of the major southern and eastern Chinese cities to other areas of China, as well as moving some of those industries to other nations as the Chinese economy moves more toward high value-added manufacturing and services. In the electricity sector, this means a combination of improved efficiency in existing and new coal fired power plants and increasing the share of power generated from non-fossil resources, including renewable energy sources (particularly hydroelectric, wind, and solar power) and nuclear power.

---

116. See, for example, Carbon Brief, "Global Historical Emissions Map", available as https://www.carbonbrief.org/interactive-map-historical-emissions-around-the-world.
117. United States Environmental Protection Agency (USEPA, 2016), "Global Greenhouse Gas Emissions Data", available as https://www.epa.gov/ghgemissions/global-greenhouse-gas-emissions-data.
118. See, for example, South China Morning Post (2017), "China orders cities to reduce emissions, as capital braces for another bout of heavy smog", dated February 13, 2017, and available as http://www.scmp.com/news/china/policies-politics/article/2070392/china-orders-cities-reduce-emissions-capital-braces.

China's nuclear sector is young by comparison to that of Japan and the ROK, but is and has been growing fast, as most of the reactors built worldwide at present are being built in China. With a large land area and a not-yet-powerful civil society sector, siting of nuclear plants and spent fuel facilities has not yet been a major problem for China, though it may grow to be so in the future.

1.2. Past and Projected Patterns of Growth of the Chinese Electricity Sector

Total electricity generation in China in 1990 was approximately 650 terawatt hours (TWh, or billion kilowatt-hours), including Hong Kong, which is about the same as present-day Germany, with about 7 percent of China's population.[119] By 2016, electricity output and consumption in China had grown nearly 10-fold, supplanting the United States in 2011 as the nation with the largest electricity consumption (see Figure 0-1). Overall generation capacity grew even more rapidly, particularly in recent years, with growth in capacity averaging over 9 percent annually from 1990 through 2005, and 11 percent annually from 2005 through 2016 (see

Figure 0-2).[120] Generation capacity in China now exceeds 1600 GW (gigawatts, or million kilowatts), nearly 60 percent more than the

119. British Petroleum (2017), *BP Statistical Review of World Energy June 2017*, excel workbook accompanying volume available at http://www.bp.com/statisticalreview.
120. Some of the even more rapid growth in the post-2005 period was due to the addition of large amounts of wind and solar power generation capacity, which have lower capacity factors, and thus generate less energy annually per unit of capacity than, for example, coal-fired and nuclear power plants.

United States, where generation capacity stood at a bit over 1000 GW as of 2016.[121]

Thermal power, and specifically coal-fired power, has been the mainstay of Chinese electricity generation. Thermal power provided about 80 percent of generation in 1990, remaining near that level though 2010, falling only in recent years to under 72 percent by 2016. Despite rapid growth in nuclear capacity, nuclear power accounted for only 3.6 percent of electricity output by 2016, somewhat less than wind power in that year. Construction of hydroelectric capacity has been rapid in the past decade and continues today with nearly 12 GWe of capacity added in 2016 alone to a total of 330 GWe (of which about 27 GWe are pumped-storage plants used for peak power provision).[122] Electricity consumption in China has been dominated by the industrial sector, which consumed nearly 77 percent of power in 1990. The importance of the industrial sector has waned somewhat—to about 70 percent of total consumption in 2014, as residential and commercial/services electricity use has grown—but still remains the major user of electricity in the Chinese economy.

The rapid overall growth in electricity consumption in the last decade, however, masks much slower growth in recent years—just over 3 percent annually between 2013 and 2016, as shown in the last four bars of

---

121. United States Department of Energy, Energy Information Administration (USDOE EIA, 2017), "Electricity in the United States", available as https://www.eia.gov/energyexplained/index.cfm?page=electricity_in_the_united_states#tab2.

122. International Hydropower Association (2017), "China", last updated May, 2017, and available as https://www.hydropower.org/country-profiles/china. This reference highlights the construction of the "Wudongde project on the Jinsha River in the south- west, which will provide 10.2 GW installed capacity when complete" in 2020, and will be the sixth-largest hydro plant in the world.

Figure 0-2—as the Chinese economy has slowed somewhat, and greater emphases have been placed on improvements in energy efficiency, development of the services sector, and the reduction of heavy industry.

Figure 0-1: Electricity Generation in China by Type, 1990 through 2016[123]

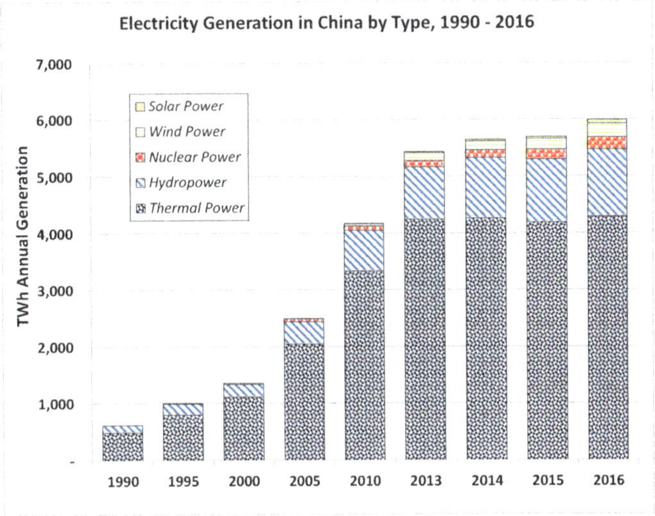

123. Sources: Figures through 2014 from *China Statistical Yearbook, 2016*, available as http://www.stats.gov.cn/tjsj/ndsj/2016/indexeh.htm, with 2015 and 2016 data from John A. Mathews and Hao Tan (2017) "China's Continuing Green Shift in the Electric Power Sector: Evidence from 2016 data", *The Asia-Pacific Journal | Japan Focus*, Volume 15, Issue 10, Number 4, May 15, 2017, available as http://apjjf.org/-John-A--Mathews--Hao-Tan/5038/article.pdf.

Figure 0-2: Electricity Generation Capacity in China by Type, 1900 through 2016[124]

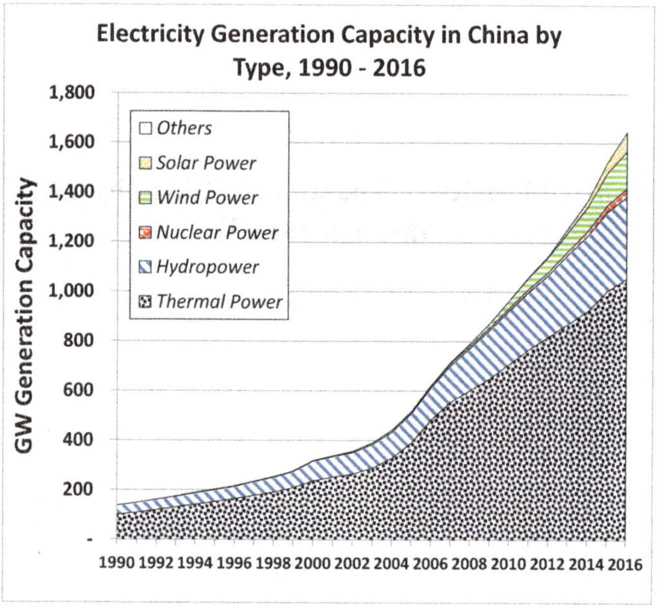

Although electricity in the 1990s and 2000s was transmitted and distributed through mostly six regional grid "clusters," recent years have seen massive investments in transmission lines designed to tie the national grid together. In addition to several point-to-point UHV DC (Ultra-high voltage Direct Current) lines, China has invested and is investing tens of billions of dollars in UHV AC (alternating current) lines.[125] Together, these UHV DC and UHV AC lines are

124. Sources of data are as indicated for Figure 1-1.

125. UHV DC lines are 800 kV (kilovolts) or more, and UHV AC lines are 1000 kV or more. By way of comparison, large high-voltage transmission lines in most countries are rated at about 500 kV. The higher the voltage, the more power can be carried by a given line. See, for example, *China Daily* (2014),

designed to move electricity from coal-fired, hydroelectric, and wind power plants in China's North and West to the major consuming cities in central and eastern China. An additional goal of these transmission projects is to reduce coal-fired power generation, with its attendant air pollution problems, in the vicinity of big coastal cities. China is reportedly investing hundreds of billions of dollars in total in electricity transmission and distribution between 2015 and 2020, doubling the 2014 length of the lines in China's transmission system to over one million kilometers.[126] Nearly half of global additions to high-voltage transmission networks during 2014 through 2020 are expected to be in China.

Massive investments in manufacturing of electricity sector equipment in China have accompanied investments in transmission and distribution infrastructure and have made China a world leader in the production of many types of power plants. China's largest wind turbine manufacturer, Goldwing, was third among global wind power firms in 2016, with an output of 6.4 GW (almost all installed in China), after leading the world the previous year.[127] China's solar

---

"China Exclusive: China to build 12 power transmission lines", dated 2014-05-14, and available as http://shanxi.chinadaily.com.cn/2014-05/14/content_17505983.htm.

126. See, for example, Reuters (2015), "UPDATE 1-China targets $300 bln power grid spend over 2015-20 – report", dated August 31, 201, and available as http://www.reuters.com/article/china-power-transmission-idUSL4N1171UP20150901, and Cision PR Newswire (2015), "Global Transmission and Distribution Market 2015 Report Says China Leading Power Transmission Line Additions to 2020", dated June 11, 2015, and available as http://www.prnewswire.com/news-releases/global-transmission-and-distribution-market-2015-report-says-china-leading-power-transmission-line-additions-to-2020-506906491.html. By way of comparison, the United States transmission grid includes about 700 thousand kilometers of high voltage lines (US Department of Energy (2014), "Top 9 Things You Didn't Know About America's Power Grid", dated November 20, 2014, and available as https://energy.gov/articles/top-9-things-you-didnt-know-about-americas-power-grid).

127. See, for example, Bloomberg New Energy Finance (2017), "Vestas reclaims top spot in annual ranking of wind turbine makers", dated February 22,

photovoltaic (PV) firms produced 71 percent of the world's PV modules in 2016—and most of their modules were installed in China, as well as exporting panels to the rest of the world.[128] Almost by definition, China leads the world in production of coal-fired power plants, installed both in China and, increasingly, in other nations.[129] And China has slowly transitioned its nuclear industry from plants built with mostly foreign technology to plants designed and built in-country.

The emphasis on manufacturing of renewable energy equipment has accompanied aggressive national goals for renewable energy deployment. Progress toward, and even past, these goals has been impressive. Deployment of solar photovoltaic power has been so rapid—topping 10 GW in a single recent month—that China's goal for solar deployment by 2020 under its 13th Five-Year Plan (FYP) has been nearly doubled, from 105 GW (already achieved) to 230 GW.[130] Progress toward wind power deployment goals has been nearly as impressive, with 129 GW of wind power capacity

---

2017, and available as https://about.bnef.com/blog/vestas-reclaims-top-spot-annual-ranking-wind-turbine-makers/.

128. Jeffrey Ball, Dan Reicher, Xiaojing Sun, and Caitlin Pollock (2017), *The New Solar System: China's Evolving Solar Industry And Its Implications for Competitive Solar Power In the United States and the World*, Stanford University Steyer-Taylor Center for Energy Policy and Finance, dated March, 2017, and available as https://www-cdn.law.stanford.edu/wp-content/uploads/2017/03/2017-03-20-Stanford-China-Report.pdf.

129. See, for example, Hiroko Tabuchi (2017), "As Beijing Joins Climate Fight, Chinese Companies Build Coal Plants", *New York Times*, July 1, 2017, available as https://www.nytimes.com/2017/07/01/climate/china-energy-companies-coal-plants-climate-change.html?mcubz=0&_r=0.

130. See, for example, Joshua S Hill (2017), "China Continues Massive Solar Installations With 10.52 GW In July, Already Exceeds 2020 Target", Cleantechnica, dated August 22nd, 2017, and available as https://cleantechnica.com/2017/08/22/china-continues-massive-solar-installations-10-52-gw-july-already-exceeds-2020-target/; and Mark Osborne (2017), "China's solar boom to continue through 2020 as install targets revised", PVTech, dated July 28, 2017, and available as https://www.pv-tech.org/news/chinas-solar-boom-to-continue-through-2020-as-install-targets-revised.

deployed by 2015, already over half of the 2020 target of 210 GW set in China's 13th FYP for energy. For some wind-rich provinces, wind and solar power already provided up to 15 percent of total generation by 2020.

China's renewable energy industries have not been without growing pains. Many wind generators, particularly in northern and western provinces, were built in anticipation of local electricity demand and/or construction of transmission facilities that have yet to catch up with wind power capacity, and as a result, wind energy output worth billions of dollars has been curtailed, far more than in other wind-rich areas of the world (such as Texas). The transmission line projects described above are expected to significantly lower curtailment rates in coming years, allowing wind energy from the North and West of China to displace coal-fired power for the cities of the East and South.[131]

Accompanying this drive to use more renewable energy has been a drive toward energy efficiency in multiple sectors. China's National 13th FYP includes a reduction of 15 percent in energy use per unit of GDP relative to the level in 2015 and a reduction of 18 percent in carbon dioxide emissions per unit of GDP.[132] It should be noted that energy (and CO_2) per unit of GDP are indicators dependent on several factors, most notably the composition of industry in China, the types of products produced, and the value of those products, as well as the actual energy efficiency per unit of physical output. CO_2 per unit of GDP additionally factors in the composition of the energy sources used by an economy. As a result, a reduction in energy use

131. See, for example, Li Ying (2016), "Blowing in the wind", Chinadialogue, dated May, 31, 2016, and available as https://www.chinadialogue.net/article/show/single/en/8965-Blowing-in-the-wind.
132. Ma Tianjie (2017), "China's Ambitious New Clean Energy Targets", *The Diplomat*, dated January 14, 2017, and available as http://thediplomat.com/2017/01/chinas-ambitious-new-clean-energy-targets/.

(and CO_2 emissions) per unit of GDP can be accomplished by a combination of true energy efficiency improvements, greater value added in products produced, offshoring of heavy (energy intensive/polluting) industries, and a shift toward production of more services, all of which are currently in play in China.

China's 13th Five-Year Plan lays out a number of goals for electricity sector development by 2020 (see Table 0-1). These include overall electricity consumption (given as a range), development of various types of generation, and other parameters. Along with the aggressive targets for renewable power development described above, the 13th FYP shows growth in electricity generation/consumption, as well as generation capacity, slowing markedly relative to experience over the past decade.

Although longer-term official forecasts of electricity demand were not available for this paper, the trend of declining growth in Chinese electricity generation and consumption is echoed and extended in a number of forecasts by other analysts. For example, in the U.S. Department of Energy's (US DOE's) *International Energy Outlook 2016*, the growth rate of electricity use in China progressively decreases from about 3.6 percent annually in 2015-2020 to 1.4 percent/yr in 2035 through 2040 (see Figure 0-3). As China's population growth will, based on the United Nations' "Medium Variant" estimate, have reached its peak just before 2030,[133] continued growth in electricity consumption late in the U.S. DOE forecast means continued growth in electricity use per person. Overall, the USDOE forecast calls for average annual growth in electricity use of just under 2.5 percent from 2012 through 2040, resulting in a doubling of 2012 electricity use in China by 2040.

---

133. Data downloaded from United Nations DESA, "World Population Prospects 2017", available from https://esa.un.org/unpd/wpp/Download/Standard/Population/.

Table 0-1: Thirteenth Five-Year Plan Main Objectives of Power Industry Development[134]

134. Rough translation of "Box 2" from [China] National Development and Reform Commission, National Energy Board, *The Thirteenth Five-Year Plan for Electricity Development (2016-2020)*, available as http://www.ndrc.gov.cn/zcfb/zcfbghwb/201612/P020161222570036010274.pdf.

Category	2015 Value	2020 Target	Annual average Growth Rate [or change]
Installed Capacity (100 million kilowatts)	15.3	20	5.50%
West to East [Transmission Capacity] (100 million kilowatts)	1.4	2.7	14.04%
Total electricity consumption (trillion kilowatt hours)	5.69	6.8 - 7.2	3.6 - 4.8%
Electricity accounted for the proportion of terminal energy consumption	25.80%	27%	[1.2%]
Per capita installed capacity (kW / person)	1.11	1.4	4.75%
Per capita electricity consumption (kWh / person)	4142	4860 - 5140	3.2 - 4.4%
Non-fossil energy consumption share	12.00%	15%	[3%]
Non - fossil generation share of installed capacity	35.00%	39%	[4%]
Conventional hydropower (100 million kilowatts)	2.97	3.4	2.80%
Pumped Storage hydro (10 thousand kilowatts)	2303	4000	11.70%
Nuclear power (100 million kilowatts)	0.27	0.58	16.50%
Wind power (100 million kilowatts)	1.31	2.1	9.90%
Solar power (100 million kilowatts)	0.42	1.1	21.20%
Fossil energy power generation installed proportion	65%	61%	[-4%]
Proportion of installed capacity as coal-fired power	59%	55%	[-4%]
Coal-fired generation capacity (hundred million kilowatts)	9	<11	4.10%
Gas-fired generation capacity (hundred million kilowatts)	0.66	1.1	10.80%
Average coal consumption of new coal-fired units (grams of standard coal / kWh)		300	
Average coal consumption of active coal-fired units (grams of standard coal / kWh)	318	310	[-8]
Transmission Line Loss Rate	6.64%	<6.5%	
Charging Facilities Construction	Meet requirements for charging 5 million electric cars		
Electricity from Alternative Sources (100 million kWh) [Presumably for Transport]		4500	

Figure 0-3: Forecast for Electricity Use in China, 2010-2040 (U.S. Department of Energy International Energy Outlook)[135]

135. Figure prepared using data from US Department of Energy's Energy Information Administration (2016), detailed tables for *International Energy Outlook 2016*, "Table: Delivered energy consumption by end-use sector and fuel; Case: Reference case | Region: China", available from

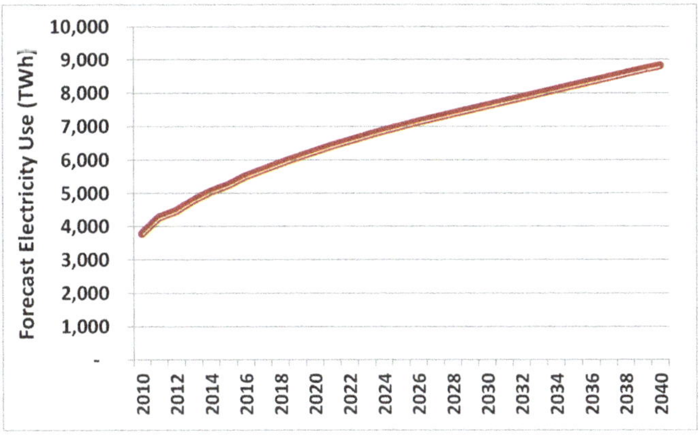

The current setting for China's nuclear power sector is thus characterized by rapid (by the standards of most countries) but slowing economic growth, increasing energy efficiency, and substantial progress on many development goals. In combination, these factors are resulting in demand for electric power that will continue to grow, but at progressively lower rates as the decades pass. At the same time, an aggressive drive to use electricity from renewable sources and reduce electricity generation (and its attendant environmental emissions) from coal creates a significant drive toward nuclear power as carbon- and air pollutant-emissions-free electricity source, but also significant competition for nuclear power among low-emissions electricity options.

1.3. China's Current and Planned Nuclear Sector: 1990s through 2040

Though the decision to develop civilian nuclear energy in China dates back to the 1970s, concrete efforts to construct nuclear power plants began only in the late 1980s. China's civilian nuclear power

---

https://www.eia.gov/outlooks/aeo/data/browser/#/?id=15-IEO2016&region=4-12&cases=Reference&start=2010&end=2040&f=A&linechart=Reference-d021916a.2-15-IEO2016.4-12&map=&sourcekey=0.

development began with the construction of the French-built reactors at Daya Bay, near Guangzhou, and providing power to Hong Kong and other cities in the region. This pair of 944 MWe reactors were constructed starting in 1987, and began operation in 1994. At about the same time, the smaller (298 MWe) Qinshan-1 unit was built with a combination of imported and domestic technology. These three reactors constituted the "first wave" of reactor construction in China, as described by the World Nuclear Association (see

Figure 0-4).[136] A second wave of reactor construction began in the late 1990s and included four additional reactors at Qinshan, near Shanghai, this time in the 600 MWe class, and produced as stepping stones to development of a "Chinese standard" 1000 MWe-class unit, the CNP-1000. Two additional units were built at the Ling Ao power plant, very close to the original Daya Bay reactors. Like the Daya Bay plant, the Ling Ao reactors are based on French technology. A pair of reactors in northern Jiangsu province, between Shanghai and Beijing, the first set of Tianwan units, were built using Russian reactor technology but with control equipment supplied by an international consortium. These first two Tianwan units were completed as the third wave of reactor construction began.

Figure 0-4: Nuclear Reactor Construction in China to Date

136. Figure and related data from World Nuclear Association (2017a), "Nuclear Power in China", updated September 2017, and available as http://www.world-nuclear.org/information-library/country-profiles/countries-a-f/china-nuclear-power.aspx.

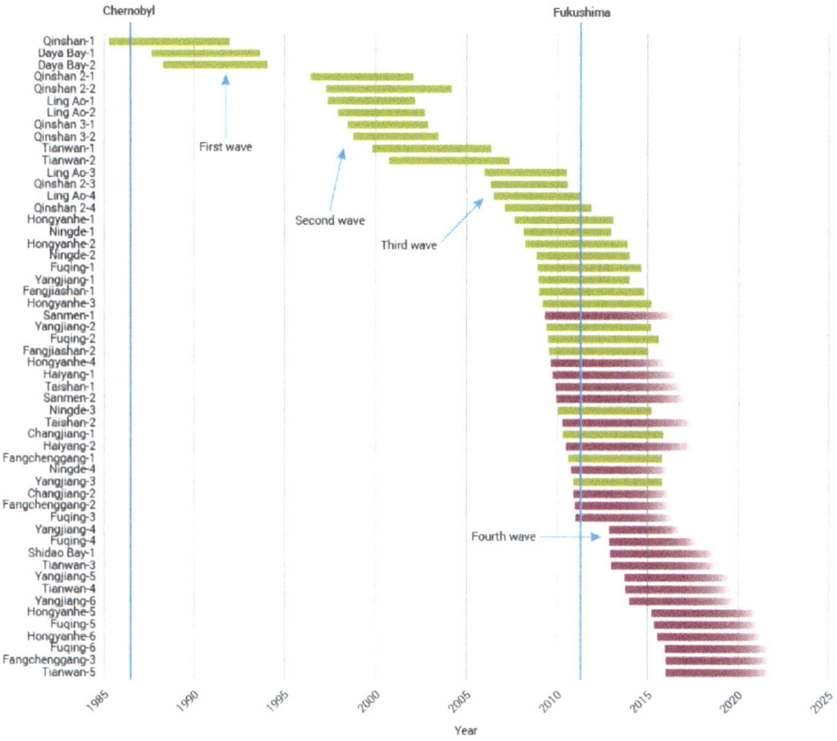

The third wave of Chinese reactor construction is considered to have begun in about 2006, with ground broken on the last few of the 30 third wave reactors just before the Fukushima accident in Japan in March of 2011. The Chinese nuclear establishment's response to the Fukushima accident was to order a delay in new reactor starts while reactor safety provisions were reviewed and strengthened, and reactor deployment plans were reviewed. Subsequent to these reviews, a fourth wave of reactor construction is considered to have begun in 2012 and 2013 and has included at least 16 reactor units to date.

1.3.1. Current Fleet of Chinese Reactors

The current fleet of operating Chinese reactors, resulting from the first three waves of reactor construction described above (the fourth wave and some of the third wave plants being not yet operational) consists of 37 units totaling 33.7 GW of electricity generating capacity. Two of these units, Qinshan Phase III, units 1 and 2, are based on the CANDU heavy water/natural uranium technology (PHWRs); all the rest of the current fleet are light water reactors using low-enriched uranium fuel.[137] As shown in Figure 0-5, all existing and under-construction reactors in China are in coastal locations, and all but three nuclear plants listed as "planned" are also in coastal locations. Plants planned for inland locations, which must use river water for cooling, have been a focus of regulatory review for Chinese authorities, particularly post-Fukushima, with the reliable availability of sufficient cooling water, particularly in a changing climate regime, being a significant concern, as well as the possible pollution of rivers in the event of an accident. China's regulatory review of inland sites may have contributed to a general impression among the Chinese public that the consequences of a nuclear accident at a reactor located at an inland site will result in radioactive contamination of the river used for cooling that would be much worse than contamination of a coastal site. This impression, to the extent that it persists, may make it more difficult for reactors to be built in non-coastal areas.

137. Light water reactors, or LWRs, use regular water (H2O) as the reactor coolant. The two primary light water reactor designs are pressurized water reactors (PWRs), which are dominant in China, and boiling water reactors (BWRs), which are also common worldwide (and of which the units at the damaged Fukushima Daiichi plant are examples). China is also building a 210 MWe high-temperature gas-cooled reactor (Model "HTR-PM") in Shandong province. The HTR-PM uses fuel spheres ("pebbles") rather than fuel encased in long metal tubes (fuel rods) like the other reactor types.

Figure 0-5: Location of Nuclear Power Plants Existing, Under Construction, and Planned in China[138]

The World Nuclear Association characterizes China's nuclear policy as:

"China has set the following points as key elements of its nuclear energy policy:

- PWRs will be the mainstream but not sole reactor type.
- Nuclear fuel assemblies are fabricated and supplied indigenously.
- Domestic manufacturing of plant and equipment will be maximized, with self-reliance in design and project management.
- International cooperation is nevertheless encouraged."

---

138. World Nuclear Association (2017a), ibid.

The World Nuclear Association also notes that "[t]he technology base for future reactors remains officially undefined, though two designs are currently predominant in construction plans: CAP1000 and Hualong One, after plans for more CPR-1000 units were scaled back post-Fukushima. Beyond them, high-temperature gas-cooled reactors and fast reactors appear to be the main priorities." For the present, China's use of many different kinds of reactors, ordered and funded by different provinces, and only loosely coordinated with power grid development, may prove to be problematic soon, and may complicate nationally coordinated management of spent fuel.[139]

The combination of current electricity generation over-capacity, particularly in east coast areas where many reactors are located, plus the variety of reactors under construction, the poor record of many imported reactor technologies, and China's ambitions to export reactors itself, combine to yield a picture of China's nuclear future that is significantly muddled relative to stated policy. The text box below, co-authored by Professor Stephen Thomas of the University of Greenwich, United Kingdom, briefly explores these issues.[140]

---

139. In addition, China's nuclear energy sector involves dozens of different firms and entities. Some of these are national in scope, some regional or provincial, and some organized specifically to implement a particular nuclear project. Ownership of nuclear entities is similarly complex, with national and provincial utilities being mostly state-owned, but with a variety of joint ventures and subsidiaries involving publicly traded companies, private entities, and others. See World Nuclear Association (2017b), "Government Structure and Ownership: Nuclear Power in China Appendix 1", updated September 2017, available as http://www.world-nuclear.org/information-library/country-profiles/countries-a-f/appendices/nuclear-power-in-china-appendix-1-government-struc.aspx.

140. For additional detail, see Steve Thomas (2017), "China's nuclear export drive: Trojan Horse or Marshall Plan?", *Energy Policy*, volume 101 (2017), pages 683–691.

Box 1: China's Nuclear Export Ambitions: Prospects and Challenges[141]

China's lack of focus on one or a very few reactor designs has resulted in a technologically challenging situation as a major reactor build-out continues. China's original plan as of 2007 was to use the 1970s/1980s French reactor design for a few years, but import and indigenize state-of-the-art foreign technologies, with the goal of making reactors affordable for domestic use and for export to countries new to nuclear power, as well as to established nuclear energy users. The indigenizing of foreign technologies was to have been done through the adaptation of the Westinghouse AP1000 and to a lesser extent the French (Areva) EPR designs. Both technologies have failed badly, so China's two major reactor vendors, CGN and CNNC, have returned to the old French technology used for, for example, the units at Daya Bay, updating it and adding new features to develop the ACP-1000 and ACPR-1000 designs, which are being "merged" to yield the Hualong One. Whether this design will be safe enough to convince foreign regulators (it is being tested in the United Kingdom, which has been considering importing Chinese reactors), and still be inexpensive enough to attract buyers, remains to be seen. Another, somewhat weaker, Chinese nuclear utility, SNPTC, which was set up to indigenize the AP1000 design, is scaling it up to the 1400 MWe

141. See also Matthew Cottee (2017), "China's nuclear export ambitions run into friction", *Financial Times*, dated August 2, 2017, and available as https://www.ft.com/content/84c25750-75da-11e7-90c0-90a9d1bc9691. Interestingly this article includes the following passage about Russia's allegedly waning interest in nuclear exports: "The current leader in the nuclear export market, Russia's Rosatom, is reportedly shifting focus to hydropower and wind turbines rather than its usual reactor business. Speaking at the 'Technoprom-2017' conference in Novosibirsk, the deputy general director of Rosatom, Vyacheslav Pershukov, suggested that the export market for nuclear reactors has been exhausted."

CAP1400 design to try to make the economics more attractive, but whether doing so will be successful or not is unclear.

Despite a substantial interest in exporting reactors, and substantial effort to do so, China has been unable to reach firm agreements except for its deal with Pakistan and an agreement, but not yet a contract, to construct (or rather, to resume construction on) two CANDU units in Romania.[142] Both of these export ventures are receiving or thought to be receiving significant Chinese financing for the purchases. A number of other deals are reportedly under discussion. The lack of additional firm contracts may be because China has only entered the field of nuclear vendors during this decade, because there is reluctance to buy from China (for reasons not completely understood) or because there is simply little actual market for reactors outside of China. The contrast between Russia, which lacks the wherewithal to provide financing and significant portions of the required the supply chain to support reactor exports, but has about 30 firm orders, and China, which does have the money and supply chain but has no orders, is marked.

The size of China makes analysis of its economic and technological situations hard to carry out, because its industries can dominate international markets without being dominant at home. Nuclear represents a negligible part of China's electricity mix, yet reactors built in China account for the majority of the world's nuclear construction over the last decade. As a result, one might say that that global nuclear industry needs China, but China doesn't necessarily need nuclear power. As such, the opposition by some citizen groups

142. See, for example, World Nuclear Association (2017c), "Nuclear Power in Romania", last updated October, 2017, and available as http://www.world-nuclear.org/information-library/country-profiles/countries-o-s/romania.aspx; and Stephen Stapczynski and Aibing Guo (2015), "China Widens Nuclear Exports With $7.7 Billion Romania Plant", *Bloomberg*, dated November 9, 2017, and available as https://www.bloomberg.com/news/articles/2015-11-10/china-keeps-nuclear-exports-going-with-7-7-billion-romania-deal.

to building nuclear power at inland sites is very important because it limits reactor development to the coastal areas, and in some coastal regions there is already significant generation overcapacity. Overcapacity in some parts of the coast has meant that reactors are being used for load-following—which A) they are not built for, and B) has a negative impact on their capacity factors and thus on their profitability—and new plants have even been delayed in entering service because there is insufficient electricity demand, further affecting the economics of nuclear investments.[143]

Given the challenges above, it is unclear to the authors how long the Chinese government will put its weight behind its nuclear export industry when the nation as a whole could win much more business and gain more political influence by putting its weight behind other technologies such as high-speed rail, in which it has had remarkable domestic success, and renewable energy, in which it has substantial and sometimes dominant market shares both at home and abroad.

1.3.2. Nuclear Fuel Cycle Facilities

China obtains portions, in some cases most, of its uranium, fuel conversion, enrichment, and fuel fabrication needs (the "front end" of the fuel cycle) from domestic resources and facilities. As a nuclear weapons state, some of these facilities were originally developed in support of China's nuclear weapons program. The World Nuclear Association describes China's policy on uranium acquisition for its nuclear power program as targeting "about one-third of uranium supply domestically, one-third from Chinese equity

---

143. See C.F. Yu (2016), "CHINA: Beijing Drafts Rule to Put Reactors In Load-Following Mode", *Nuclear Intelligence Weekly*, Volume 10, No. 33, August 19, 2016, available from http://www.energyintel.com/pages/about_uiw.aspx.

in foreign mines, and one-third on the open market."[144] China has seven operating uranium mines, two of which date from the 1960s and 70s, as well as uranium resources in many other locations. China has purchased a significant amount of uranium from a number of supplier nations, and two Chinese firms hold equity in uranium mines in central Asia, Africa, and Canada.

Enrichment of uranium is the process of concentrating the fraction of the U-235 isotope found in natural uranium (about 0.7 percent) to the fraction needed in LWR fuel, typically 3 to 5 percent. A preliminary step to enrichment is conversion of natural uranium oxide, a solid, to gaseous uranium hexafluoride (UF_6). The World Nuclear Association lists China's conversion capacity as somewhat uncertain, with a report of a 5000 t/yr plant in Gansu province operating at 80 percent of capacity, with another 9000 t/yr plant due to come on line this year or next. A smaller plant (500 tU/yr) plant is reported in northwest Gansu province. Another plant with a capacity of 3000 t/yr is reportedly being built by China Nuclear Fuel Corp in Hunan province, and will be on line in 2018.

Enrichment of uranium in China takes place at larger plants in Shanxi and Gansu provinces, and at two smaller plants in Sichuan province. The total capacity of these plants is estimated at 5.7 to 7.0 million SWU (separative work units, a measure of enrichment capacity) per year in 2015, and a projected range of 10.7 to 12.0 million SWU/yr in 2020. By way of comparison, annual enrichment demand for Chinese reactors was expected to total 9 million SWU in 2020. Enriched uranium has also been imported to China, particularly for reactors of foreign design, with enrichment services provided in Europe and Russia. Centrifuges used for enrichment were provided by Russia in the past, but indigenous centrifuges

144. World Nuclear Association (2017b), "China's Nuclear Fuel Cycle", updated July 2017, and available as http://www.world-nuclear.org/information-library/country-profiles/countries-a-f/china-nuclear-fuel-cycle.aspx.

technology has been used for recent capacity expansions starting in 2010.[145]

The fabrication of fuel for Chinese reactors is done mostly in China, with some reactors supplied by France and Russia under contract to receive fuel from those nations sufficient for the first reactor core loading and a number of subsequent loadings. A plant in Yibin, Sichuan province, with a total capacity of 900 tU/yr, makes fuel for PWRs and for the Russian VVER design. A second major plant in Inner Mongolia, at Baotou, fabricates fuel for China's pair of CANDU reactors as well as for various PWR models[146] and will reportedly have total annual capacity of 1600 tU/yr by 2020. A facility at the Baotou plant is also making fuel for the high-temperature gas reactor being completed in Shandong province. Fuel pellets for the fuel assemblies for some reactor models are also sourced from Kazakhstan's Ulba Metallurgical Plant.

Most spent PWR fuel in China is currently stored in pools on reactor sites. Generally, the on-site spent fuel storage capacity at operational nuclear power plants can accommodate 10 years of spent fuel. Taking into account ongoing trends in nuclear fuel management, such as increasingly high rates of fuel burnup (which reduce the number of refueling cycles necessary), extensions of reload cycles, and the use of dense-pack storage in spent fuel pools into consideration, it is estimated that the storage capacity of present facilities can be enlarged to hold approximately 20 years' worth of spent fuel. Currently, all PWR spent fuel is in fact stored at reactor

145. Hui Zhang (2015), *China's Uranium Enrichment Capacity: Rapid Expansion to Meet Commercial Needs*, Project on Managing the Atom, Discussion Paper No. 2015-03, Belfer Center for Science and International Affairs, Harvard Kennedy School, dated August, 2015, and available as http://www.belfercenter.org/sites/default/files/legacy/files/chinasuraniumenrichmentcapacity.pdf.
146. Including the AP1000 and its Chinese variant the CAP1000, based on Westinghouse designs and intended to be the "main basis of China's move to Generation III technology" (see World Nuclear Association 2017a, ibid).

sites except for some of the spent fuel removed from the Daya Bay reactors, the first units in commercial operation in China. Since 2003, shipments of spent fuel from Daya Bay have been transported approximately twice annually to the centralized interim storage facility in Gansu province, where they are placed in wet storage facilities (away-from-reactor spent fuel pools).[147]

The facility in Gansu province, the Jiuquan Atomic Energy Complex (JAEC) as noted above, was initially developed in the 1950s and 60s to support China's nuclear weapons program. Since then, facilities for storage of civilian nuclear spent fuel have been added, as well as a pilot-scale reprocessing plant, constructed starting in 1986 but not operational until 2010, that can handle 60 tons of spent nuclear fuel (expressed as the mass of heavy metal—mostly uranium plus plutonium—abbreviated as tHM) annually.[148]

1.3.3. Plans and Projections for Reactor Deployment

The World Nuclear Association lists a total of 20 reactors in China as "under construction" and 30 to 40 more as "planned," mostly with an expected construction start before 2019. The under construction and planned units total 66.7 GWe.[149] In addition, the same reference lists 100 units, totaling 114.2 GWe, as "proposed," and an additional 79 units, and 90.8 GWe, proposed for the more distant future.

---

147. Liu Xuegang (2014), *Spent Nuclear Fuel Management in China*, Nautilus Institute NAPSNet Special Report, dated 5 August, 2014, and available as https://nautilus.org/napsnet/napsnet-special-reports/spent-nuclear-fuel-management-in-china/.
148. See, for example, Nuclear Threat Initiative (2011), "Jiuquan Atomic Energy Complex", last updated September 29, 2011 , and available as http://www.nti.org/learn/facilities/722/.
149. World Nuclear Association (2017a), ibid. For nuclear power plants, capacity is sometimes shown as GWe, or gigawatts of electric power, to differentiate from output expressed in terms of thermal power, which is typically about three times higher.

Together, these listings plus the reactors already in operation total well over 300 GWe, over three times the size of the United States reactor fleet at its historic maximum in 2012.[150]

One recent article, citing both the World Nuclear Association source referenced above and a Chinese-language study by the Chinese Academy of Engineering lists ranges of nuclear capacities of 83.8 to 200 GWe by 2030, and 250 to 400 GWe by 2050.[151] Other reported projections provide ranges of 340 to 500 GWe by 2050,[152] 1200 TWh of generation by 2040,[153] and on the order of 250-300 GWe, by 2050, the latter in a "two-degree" scenario where fossil fuel use is largely phased out except with the use of carbon capture and storage technologies.[154] These projections for massive growth in nuclear capacity are tempered by reports identifying the significant challenges to the Chinese nuclear sector in the years to come,[155] and

---

150. Nuclear Energy Institute (2017), "US Nuclear Generating Statistics, 1971 – 2016", updated April, 2017, and available as https://www.nei.org/Knowledge-Center/Nuclear-Statistics/US-Nuclear-Power-Plants/US-Nuclear-Generating-Statistics.
151. Qiang Yue, Jingke He, Laurence Stamford, and Adisa Azapagic (2016), "Nuclear Power in China: An Analysis of the Current and Near-Future Uranium Flows", *Energy Technology*, 2017, Volume 5, Issue 5, pages 681 – 691, available as http://onlinelibrary.wiley.com/doi/10.1002/ente.201600444/epdf.
152. Brian Wang (2015), "China will complete five nuclear reactors in 2017 and double nuclear power generation to about 420 TWh by 2021", NextBigFuture, dated March 2, 2017, available as https://www.nextbigfuture.com/2017/03/china-will-complete-five-nuclear.html.
153. US Department of Energy, Energy Information Administration (2016), "China expected to account for more than half of world growth in nuclear power through 2040", dated September 28, 2016, and available as https://www.eia.gov/todayinenergy/detail.php?id=28132, 1200 TWh at an average annual capacity factor of 80 percent would translate into about 170 GWe of nuclear generation capacity.
154. International Energy Agency/Nuclear Energy Agency (2015), *Technology Roadmaps: Nuclear Energy, 2015 Edition*, available as https://www.iea.org/media/freepublications/technologyroadmaps/TechnologyRoadmapNuclearEnergy.pdf.
155. See, for example, Xu Yi-chong (2016), "China's contested nuclear future: The expansion of China's nuclear power production faces some serious challenges", Asia Pacific Policy Forum, dated 5 February 2016, and available as

focusing on renewable power and efficiency as alternatives to both fossil- and nuclear generation.[156]

1.3.4. Plans for Future Nuclear Fuel Cycle Facilities

The World Nuclear Association has summarized China's known plant for future development of "front-end" (nuclear fuel supply) and "back-end" (spent fuel management) fuel cycle facilities.[157] China looks to expand a number of mining activities, including, interestingly, extraction of uranium from coal ash from a power station in Yunnan province. Even with such expansions, based on at least one analysis,[158] it seems unlikely that China will be able to produce one-third of its uranium needs domestically, thus the significant emphasis by Chinese firms on investments in and joint ventures with companies prospecting for and producing uranium abroad.

Hui Zhang of Harvard University's Project on Managing the Atom, in a 2015 report, suggests, based on interviews with Chinese experts,

---

https://www.policyforum.net/chinas-contested-nuclear-future/. This article cites challenges to nuclear build-out including difficulties in reactor siting, the use of multiple technologies as an impediment to cost reduction and regulation, other challenges in organizing and making effective the regulation of the industry, and competitive realities associated with the slowdown in the growth of electricity demand coupled with the growth of deployment of and ongoing cost reduction in the renewable power industries.

156. Energy Research Institute, Lawrence Berkeley National Laboratory, and Rocky Mountain Institute (2016), *Reinventing Fire: China, Executive Summary* dated September 2016, and available as https://www.rmi.org/wp-content/uploads/2017/05/OCS_Report_ReinventingFireChina_2016.pdf. Graphics in this document suggest that under a "Reinventing Fire" scenario, China's electricity generation from nuclear power in 2050 would be about 1300-1400 TWh, or the equivalent of about 190 GWe of capacity at an average capacity factor of 80 percent.

157. World Nuclear Association (2017b), ibid.

158. Qiang Yue, Jingke He, Laurence Stamford, and Adisa Azapagic (2016), ibid.

that Chinese capacity to expand its enrichment capacity in the late 2010s was about 1 million SWU per year.[159] Given existing capacity, ongoing expansion, and the importation of initial cores for several reactors built with imported technology, Zhang estimates that China will easily have sufficient enrichment capacity to meet domestic needs through 2020, and perhaps will be able to sell surplus enrichment services internationally, "consistent with the government's stated policy of 'self-sufficiency' and 'targeting the international markets' in the supply of enrichment services." Information on China's plans for expansion of enrichment capacity beyond 2020 was not immediately available, but it seems likely that continued expansion of domestic facilities, coupled with the availability of capacity internationally, will be sufficient to fuel China's reactors on an ongoing basis.

Data provided by the World Nuclear Association suggests that China will have fuel fabrication facilities for PWRs of about 2400 tU/yr by 2020, sufficient to meet demand by that year. Some fabricated fuel is imported, particularly, as noted above, for the new cores of reactors built with imported technology. Information on plans for expansion of fuel fabrication facilities, as with enrichment, was not immediately available, but it is assumed that the China National Nuclear Corporation (CNNC), as the entity responsible for fuel fabrication in China, will continue to expand capacity to meet domestic demand, and possibly some fuel exports as well.

Reports indicate at least two ongoing efforts by China to build new facilities for reprocessing spent PWR fuel to separate plutonium for use in mixed-oxide fuel and, eventually, for use in fast reactors. A "medium scale" facility with a capacity of 200 tHM/yr is to be built at a site in Gansu province about 100 km away from the existing pilot-scale reprocessing plant described above. The medium-scale plant is part of an overall project of CNNC called "Long Teng 2020"

159. Zhang (2015), ibid.

(Dragon Soars 2020). The overall project was reported to have an expected cost 100 billion RMB (about USD 16 billion), [160] of which the reprocessing plant is an unspecified fraction, although an independent estimate provides a range of USD 3.2 to 5.7 billion (capital costs only).[161] In addition, CNNC is also negotiating with the multinational nuclear fuel cycle vendor Areva to develop an 800 tHM/yr reprocessing facility at a site on China's east coast. At 800 tHM/yr, this facility would be the size of Japan's Rokkasho reprocessing facility.

Overall, as reported by the World Nuclear Association, China is researching (in a program dating back to 1964) and plans use of "recycled" fuel on three tracks, as summarized in Figure 0-6. First, mixed-oxide fuel (MOX), which blends uranium with plutonium recovered from spent fuel during reprocessing, will be used in existing and new PWRs. Second, MOX will ultimately be used for fuel a new set of Fast Neutron, or Fast Breeder Reactors (FBRs in the Figure below). A small test fast reactor, the Chinese Experimental Fast Reactor, was built near Beijing was with Russian collaboration and started operation in 2011. A larger test fast reactor (600 MWe) is scheduled to be constructed stating in December 2017, with operation in 2023. Assuming a positive decision for deployment in 2020, China's first commercial fast reactor (the Chinese Commercial Fast Reactor, listed at 1000 to 1200 MWe) will be constructed starting in 2028, with commissioning in 2034. Fast reactor capacity of 40 GWe is "envisaged" by 2050. Third, a blend

---

160. Hui Zhang (2015), "China is said to be building a demonstration commercial reprocessing plant", International Panel on Fissile Materials, dated September 23, 2015, and available as http://fissilematerials.org/blog/2015/09/china_is_said_to_be_build.html.
161. Matthew Bunn, Hui Zhang, and Li Kang (2016), *The Cost of Reprocessing in China*, Project on Managing the Atom Report, Belfer Center for Science and International Affairs, Harvard Kennedy School, dated January, 2016, and available as http://www.belfercenter.org/sites/default/files/legacy/files/The%20Cost%20of%20Reprocessing.pdf.

of plutonium and recycled uranium from reprocessing of PWR fuel, plus and depleted uranium from enrichment, will be used in pressurized heavy water reactors (CANDU units, or PHWRs). A test of the use of blended recycled fuels in unit 1 of the Qinshan Phase III PHWRs was carried out, and deployment of the recycled fuels in both PHWR units was planned for 2018.

Figure 0-6: Nuclear Fuel Cycles Planned for China[162]

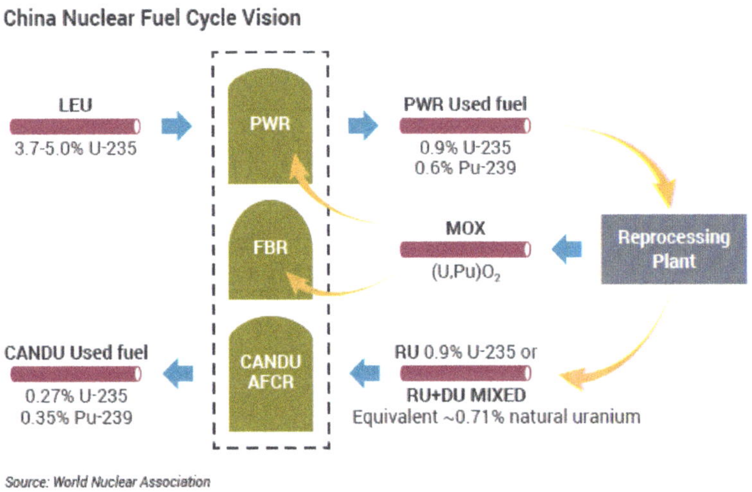

Apart from reprocessing, with regard to spent fuel management in China, a report on spent fuel storage by Professor Liu Xuegang concluded, in part:[163]

"Dry storage is currently only used for CANDU reactors in China, and will be implemented for HTR spent fuel. These two reactor models account for only a minor portion of the whole Chinese nuclear fleet. But the utilization of dry storage and its performance

162. World Nuclear Association (2017b), ibid.
163. Liu Xuegang (2014), ibid.

will have a great impact on future decision-making for the sector. Though some experts consider that the pool capacity at reactors in China is large enough to accommodate spent fuel for the next 5 to 10 years, there are strong voices supporting the building of a large-scale centralized spent fuel storage facility soon. In part, it is argued, the current practice of pool storage for spent fuel in highly dense packed arrays has been subject to criticism following the Fukushima accident. In case dense-racking is ultimately not chosen as a means of spent fuel storage in China, the decrease in potential spent fuel storage density will result in a lack of storage space at Chinese reactors relatively soon. For at-reactor storage, it is difficult to build new pools to store spent fuel due to the complexity of the pool systems. Dry storage is very promising in those cases. For centralized storage away from reactors, dry storage is still a strong competitor to pool-type storage due to advantages such as low investment, modular design, and easy maintenance. As a result, though dry storage has not been adopted for PWR spent fuel storage in China, the utilization of dry storage facilities is a strong possibility in the short or medium-term."

In essence, it is clear that China is still keeping multiple options open, from once-through fuel cycles with medium-term spent fuel storage in dry casks to fast reactor options with reprocessing of spent PWR fuel, and authorities have not yet converged on a single path for the nuclear sector. But as spent fuel builds up at existing reactors, reaching decisions regarding the future management of nuclear spent fuel will become more imperative.

2. Summary of Business-As-Usual/Baseline/Reference Electricity Sector Scenario for China

2.1. Introduction

The current status of and recent trends in China's electricity sector, as described above, forms the basis from which the sector, and the nuclear energy components thereof, will evolve in the future. Although many types of "disruptive" events can occur to suddenly change how the electricity system may evolve—an accident at (as in Fukushima) or attack on a nuclear facility being a prime, but hardly exclusive, example—in general projections of electricity futures assume continuations of existing trends, informed by expected changes in demographics, economic development, regional, national, and international policies, and other "drivers". To systematically compare policy-driven scenarios for electricity sector development, it is important to start with a "reference" case for development of the sector (and the economy in which it is embedded) to provide a consistent basis for comparison of both qualitative and quantitative attributes between scenarios. Such a reference, or "business as usual" (BAU) case is described briefly below, prepared as a composite of several studies of China's electricity future. This BAU case serves as the basis for exploration of the alternative scenarios presented in Section 3 of this paper.

2.2. Description of Composite BAU Scenario

The BAU (or reference) scenario presented here represents a composite of a number of literature sources, including work by the U.S. DOE Energy Information Administration, British Petroleum, Lawrence Berkeley National Laboratory, and others as described in Section 3. The focus is on the future of electricity generation in China, and in particular, nuclear energy's role in same, but more general economy-wide metrics are provided as points of reference and comparison with other studies.

Some of the key general parameters of the BAU Scenario for China are as follows:

- Reflecting the maturing Chinese economy, gross domestic product (GDP) growth slows over time, averaging 5.8 percent annually from 2015 through 2020 and 5.4%/yr from 2020 through 2030, but slowing to 3.5%/yr from 2030 through 2050 and 2.5%/yr from 2040 through 2050.[164]

- As projected by the United Nations, China's population growth slows through 2030, when it peaks at about 1.45 billion, declining thereafter to about 1.37 billion by 2050.[165]

- Growth in electricity generation also slows, continuing recent trends. Generation (and thus consumption, which is equal to generation less transmission and distribution losses) growth falls from 6.8 percent annually in the first half of the 2010s to 3.5%/yr from 2015 through 2020, then to 1.75%/yr from 2020 through 2030, 1.3%/yr from 2030 through 2040,

---

164. Projections shown here are consistent with those included in the US DOE Energy Information Administration (USDOE EIA, 2017) *International Energy Outlook, 2017*, as included in the downloaded table "World gross domestic product (GDP) by region expressed in market exchange rates", available from https://www.eia.gov/outlooks/aeo/data/browser/#/?id=4-IEO2016&cases=Reference&sourcekey=0. Note that GDP growth values are not used as direct drivers of electricity consumption in the BAU scenario presented here, but are provided for reference.

165. United Nations Department of Economic and Social Affairs, Population Division (2017), *World Population Prospects: The 2017 Revision*, dated June, 2017. Data shown are based on data for China in table "Medium fertility variant, 2015 – 2100". Downloaded File WPP2017_POP_F01_1_TOTAL_POPULATION_BOTH_SEXES.xlsx.

and 0.8 percent annually thereafter through 2050.[166] This trend is another reflection of a maturing economy and, coupled with continuing (though declining) GDP growth, results in falling energy intensity per unit of GDP. The declining population after 2030, however, means that growth in electricity production (and use) per person grows faster than overall electricity use, increasing, for example, at 1.5%/yr from 2030 through 2040 and 1.2%/yr from 2040 through 2050.

- While total electricity generation nearly doubles, from about 5700 TWh in 2015 to over 10,000 TWh by 2050, shares of electricity generation in China continue to shift over time, as shown in Figure 0-1. The share of electricity output provided by coal-fired power declines from about 70 percent in 2015 to slightly over 40 percent by 2050, with growth in gas-fired, wind, and nuclear generation providing nearly all of the displacement. Hydroelectric generation, as a share of the total, changes relatively little over time, solar grows substantially but to only 4.1 percent of generation by 2050, and the use of liquid petroleum products, never a large fraction of generation in China, continues to decline. Nuclear power's share of generation grows to 13 percent by 2050, but most of the growth in its share of the power market occurs before 2030.

- Changes in generation capacity by type, as shown in Figure 0-2, reflect the same general trends as in electricity generation itself, except that wind and solar power, due to their lower capacity factors (operating fewer hours per year due the intermittent availability of wind and solar resources)

166. These increases in electricity use are generally consistent with the BAU scenario for China included the USDOE EIA *International Energy Outlook, 2017* document, as referenced above.

relative to coal-fired and nuclear units, account for a larger share of added and total generation capacity. Natural-gas-fired power undergoes a shift from being mostly a peak resource, with low capacity factors (23 percent in 2015 and 2020) to being mostly a baseload resource (capacity factor of over 60 percent in 2050), reflecting a shift from simple-cycle gas turbines to more efficient combined-cycle plants.

- Total nuclear generation capacity in the BAU case rises to about 140 GW in 2040 and 170 GW in 2050. As such, it assumes a slightly reduced growth trend, particularly after 2040, than the BAU projections for capacity provided in the USDOE's most recent *International Energy Outlook*, which calls for 139 GW of generation in 2040, and 187 GW in 2050.

Additional details on the nuclear fuel cycle elements of the BAU scenario are provided in section 3, below.

Figure 0-1:

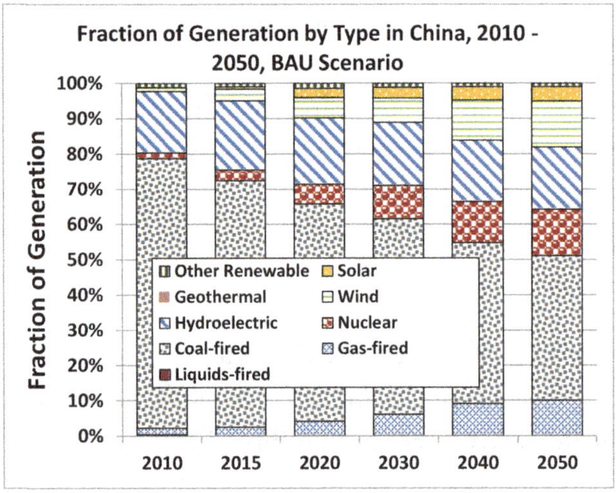

Figure 0-2:

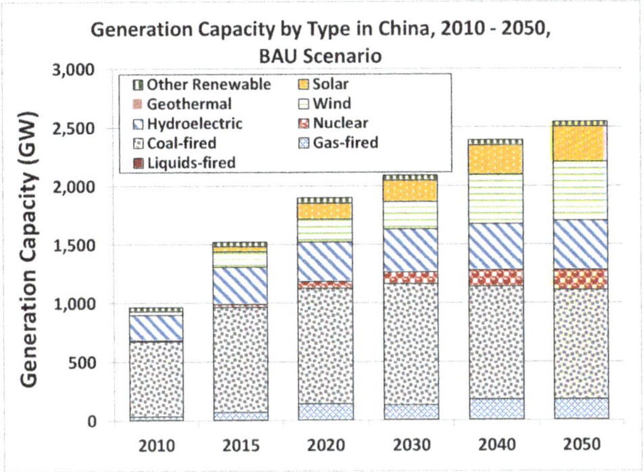

3. Alternative Scenarios for China's Energy and Nuclear Future

3.1. Introduction—Focus of Alternative Scenarios and Key Examples

Alternative scenarios, or alternative projections (which are not necessarily the same thing), can be prepared to demonstrate different potential energy futures for a nation, state, province, city, region, or other jurisdiction. Scenarios and projections are thus tools that analysts, and the policymakers they serve or seek to influence, use to test how policies devised and implemented in the near or more distant future are likely to affect key metrics that are of concern to policymakers and to society as a whole. These metrics will typically include quantitative measures such as total energy or electricity use, total cost, and emissions of local and greenhouse gas pollutants, but may also include qualitative metrics such as the expected impact on energy or military security, environmental or political risk, or social impacts.

As there are literally an infinite number of future scenarios and projections that can be chosen, it is important for analysts to select examples for comparison that are plausible—though looking out more than 30 years into the future is by definition an exercise in speculation—yet are sufficiently different that the comparison of the different cases yield policy-relevant insights. At the same time, the different scenarios/projections should typically be configured to provide the same *energy services*, that is, for example, to support economies of roughly the same size and growth rate, to move the same number of people the same distance (if not always by the same modes), and to heat and cool homes to the same degree, though not always with the same energy sources. For the different China energy futures we describe below—focusing on electricity—we have defined scenarios/projections that support approximately the same economic structure and GDP growth rates and the same populations, but are quite different in terms of how the energy systems, and especially the electricity and nuclear energy sectors, might evolve.

Given the global importance of the Chinese energy sector, a number of different groups have, over the past several years, prepared their own versions of how energy supply and demand in China might evolve through 2040 or 2050. Some notable examples include:

- The U.S. Department of Energy's Energy Information Administration projections as a part of their *International Energy Outlook* series, the 2016 and 2017 versions of which have been used to help define the BAU case projections for the Chinese electricity sector presented in section 2 of this paper.

- Work by the China Energy Group of Lawrence Berkeley National Laboratory (LBNL), including studies developing and analyzing scenarios for the evolution of China's energy sector through 2050.

- Work by the Rocky Mountain Institute and LBNL under the "Reinventing Fire" project, which looks at scenarios to vastly reduce energy sector greenhouse gas emissions in several countries, including China.

- A study of scenarios of the Chinese power sector by a group from the Renewable and Appropriate Energy Laboratory of the University of California-Berkeley.

- Work by Greenpeace, including on the future of nuclear and coal-fired power in China, and scenarios of accelerated deployment of renewable energy systems.

- Scenarios of the evolution of the energy sector published by the Energy Research Institute (ERI) of the National Development and Reform Commission (NDRC).

- Continuing work under the China Renewable Energy Outlook project, implemented by ERI and the China National Renewable Energy Centre (CNREC), with cooperation from the Danish Energy Agency and the U.S. National Renewable Energy Laboratory (NREL).

Brief summaries of each of these efforts are presented briefly below, followed by quantitative and qualitative comparisons of three "composite" scenarios that draw from these studies and others, and feature deployment of nuclear energy at different levels.

3.2. Summary of Examples of Alternative Scenarios

Alternative scenarios explore and project futures for China's energy sector that are different than a reference or business-as-usual case, and thus reflect the application of policies designed to steer the energy sector in a given direction, the influence of changes in the energy sector (or broader economy and society) not anticipated in the reference case, or both. Summaries of several alternative energy and, especially, electricity-sector scenarios for China are provided below. Most of these scenarios focus primarily on accelerated implementation of energy efficiency and/or renewable energy, relative to a reference case, but each could or would have significant implications for the nuclear energy sector as well.

3.2.1. Lawrence Berkeley National Laboratory (LBNL) China Energy Group

The researchers in the China Energy Group of the Energy Analysis and Environmental Impacts Division of LBNL have carried out a number of studies in which they have developed scenarios of China's energy future, usually focusing on energy efficiency and/or reducing greenhouse gas emissions.[167] A recent (2017) publication by the China Energy Group compares a "reference case" with alternative cases for electricity demand. A "Cost-effective Efficiency and Renewables Scenario" reduced year-2050 electricity demand by 21 percent, relative to the reference case, and a "Maximum Electrification Scenario"—based on the Cost-effective Efficiency and Renewables Scenario, but incorporating additional electrification in all sectors— still reduced 2050 electricity demand by 13 percent. Both of these scenarios resulted in a reduction of over

---

167 See, for example, Nina Khanna, David Fridley, Nan Zhou, Nihan Karali, Jingjing Zhang, and Wei Feng (2017), "China's Trajectories beyond Efficiency: CO2 Implications of Maximizing Electrification and Renewable Resources through 2050", dated May, 2017, prepared for the European Council for an Energy Efficient Economy's 2017 Summer Study, and available as https://china.lbl.gov/sites/default/files/1-142-17_zheng_khanna_final.pdf. The China Energy Group uses the LEAP (Long-range Energy Alternatives Planning) software tool developed by the Stockholm Environment Institute—United States to quantitatively model energy flows, emissions, and other parameters for China in LBNL's China 2050 Demand Resources Energy Analysis Model (DREAM).

50 percent in national CO_2 emissions in year 2050 relative to the reference case (from 11.57 billion tonnes, or Gt, CO_2 to 4.79 and 4.72 Gt CO_2, respectively). A third alternative scenario, including maximum deployment of demand-side renewables, offered additional reductions, to 3.98 Gt CO_2 by 2050. By way of comparison, China's 2016 CO_2 emissions stood at about 10 Gt, so each of the alternative scenarios represents a significant reduction, by 2050, relative to current emissions.[168]

In a 2016 study focused on energy demand and CO2 emissions in China's cities—which produce on the order of 75-80 percent of China's national CO2 emissions in recent years—LBNL China Energy Group researchers evaluated a "Low Carbon" scenario representing "a pathway in which commercially-available cost-effective efficiency and renewable energy technologies are fully deployed." [169] The measures included in the Low Carbon scenario reduced year 2050 reference case urban emissions by approximately two-thirds, a result consistent with the findings in the 2017 study detailed above.

In a 2016 study focusing on the implications of power-sector policies on coal-fired generation and CO_2 emissions in China, the China Energy Group projected that a "[s]trengthened renewable

---

168. See, for example, Zhu Liu (2016), *China's Carbon Emissions Report 2016: Regional Carbon Emissions and the Implication for China's Low Carbon Development*, Environment & Natural Resources Program, Belfer Center for Science and International Affairs, Harvard Kennedy School, dated October, 2016, and available as https://www.belfercenter.org/sites/default/files/files/publication/China%20Carbon%20Emissions%202016%20final%20web.pdf; and David Fridley, Hongyou Lu, and Xu Liu (2017), *Key China Energy Statistics 2016*, available as https://china.lbl.gov/sites/default/files/ced-9-2017-final.pdf.

169. Nina Z. Khanna, David Fridley, Lynn Price, Nan Zhou, and Stephanie Ohshita (2016), "Estimating China's Urban Energy Demand and CO2 Emissions: A Bottom-up Modeling Perspective", 2016 ACEEE Summer Study on Energy Efficiency in Buildings, available as https://china.lbl.gov/sites/default/files/11_129.pdf,

MMS [mandatory market share] with green dispatch scenario" could reduce coal-fired generation to 14 percent of total generation by 2050.[170] In the same scenario, renewable power generation (biomass, wind, solar, and hydro) provides about 63 percent of generation in 2050, with output of "renewables" (inferred in this case to mean wind, solar, and biomass) reported at 4472 TWh, and nuclear power providing about 20 percent of total generation. At that level of nuclear generation, the implied capacity of nuclear power assumed in the strengthened renewable MMS with green dispatch scenario would be nearly 300 GW.[171]

3.2.2. Rocky Mountain Institute, LBNL, and Energy Research Institute "Reinventing Fire" Project

An ongoing collaboration between the Rocky Mountain Institute (RMI), LBNL, and the Energy Research Institute (ERI) of the of the National Development and Reform Commission of China (see below), the "Reinventing Fire" project seeks to identify pathways of deep carbon dioxide emissions reduction through a combination of energy efficiency, electrification of end-uses that currently use fossil fuels, and expanded deployment of renewable electricity. The collaboration draws on work by all three partners, including the LBNL and ERI described above and below, respectively. An Executive Summary of the *Reinventing Fire* report from 2016 shows

---

170. Nina Zheng Khanna, Nan Zhou, David Fridley, and Jing Ke (2016), "Quantifying the potential impacts of China's power-sector policies on coal input and CO2 emissions through 2050: A bottom-up perspective", *Utilities Policy*,
Volume 41, August 2016, Pages 128-138, available as available from https://eta.lbl.gov/publications/quantifying-potential-impacts-chinas.
171. Estimate of implied nuclear capacity by the author of this paper is based on review of Figure 3 in Khanna et al, 2016 (ibid), and assuming an average capacity factor for nuclear generation in 2050 of 85 percent. Reference case generation in 2050 was reported as 10,730 TWh.

a reference case in which electricity generation rises to 12,800 TWh by 2050, up from 4,700 TWh in a 2011 base year.[172] In the "Reinventing Fire" case, electricity generation in 2050 falls by 16 percent relative to the reference scenario, to 10,800 TWh, as a result of two partially offsetting sets of policies. Energy efficiency reduces 2050 electricity demand by on the order of 6,000 TWh, relative to the reference case, while electrification displacing fossil fuel use throughout the economy adds back 4,000 TWh to the 2050 total. Additional renewable electricity generation in the Reinventing Fire scenario displaces about 3,000 TWh of fossil fuels and some nuclear generation in 2050, relative to the reference case. Nuclear generation in 2050 in the Reinventing Fire case is about 1,550 TWh, as opposed to about 2,400 TWh in the reference case. These values correspond to 2050 nuclear capacities of about 210 and 320 GW, respectively, assuming an average capacity factor for nuclear generation of 85 percent in 2050.[173]

Real GDP growth in both cases was assumed to be slightly higher than that included in the BAU case outlined in Section 2.2, above, at annual averages of 7.18 percent from 2010-2020, 5.60 percent from 2020-2030, 4.12 percent from 2030-2040, and 2.94 percent from 2040-2050. Population growth assumptions were similar to those in the United Nations "Medium Variant" projections reported above.

The net result of the Reinventing Fire scenario is to essentially cut in half year 2050 primary energy requirements relative to the reference case, bringing overall primary energy use by the Chinese economy back to 2050 levels. Overall 2050 CO_2 emissions similarly

---

172. ERI, LBNL, and RMI (2016), *Reinventing Fire: China, A Roadmap For China's Revolution in Energy Consumption and Production To 2050, Executive Summary*, dated September, 2016, and available from https://www.rmi.org/insights/reports/reinventing-fire-china/.
173. Nuclear generation values estimated by the author of this paper from data presented in Figure ES-14 in ERI, LBNL, and RMI (2016), ibid.

fall by nearly half in the Reinventing Fire case, relative to the reference case. Emissions of the local and regional air pollutants sulfur and nitrogen oxides fall by factors of approximately 8 and 12, respectively, relative to 2010 level, in the Reinventing Fire scenario. Further, the authors of the report find a net direct economic benefit due to the transition to the Reinventing Fire scenario. The Reinventing Fire Scenario will

> "...save China 21 trillion RMB in energy costs. From 2010 to 2050, implementing the Reinventing Fire Scenario yields a potential energy savings of 56 trillion RMB ($8.3 trillion) relative to the Reference Scenario. Incremental new investment required beyond the Reference Scenario to realize these energy savings is estimated to be 35 trillion RMB ($5.2 trillion), yielding a net present value savings of 21 trillion RMB ($3.1 trillion, all figures 2010 real)."

3.2.3. "SWITCH-China" Modeling by UC Berkeley Renewable and Appropriate Energy Laboratory

SWITCH-China is an integrated model of the Chinese electricity sector prepared by researchers at the University of California's Renewable and Appropriate Energy Laboratory (RAEL) and their colleagues.[174] In a 2016 study, the SWITCH-China model was used to investigate four scenarios of the evolution of carbon emissions by the Chinese power sector. The most stringent of these, the "IPCC Scenario," includes on the order of 250-300 GWe of nuclear

174. Gang He, Anne-Perrine Avrin, James H. Nelson, Josiah Johnston, Ana Mileva, Jianwei Tian, and Daniel M. Kammen (2016), "SWITCH-China: A Systems Approach to Decarbonizing China's Power System", *Environmental Science and Technology,* 2016, 50 (11), pp 5467–5473, available as http://pubs.acs.org/doi/abs/10.1021/acs.est.6b01345.

generation capacity by 2050,[175] meeting 14 percent of total electricity demand, along with about 1500 GWe of wind power, about 1900 GWe of solar power, and 500 GWe of electricity storage. By 2050, about 90 percent of coal-fired generation (providing in total 29 percent of China's 2050 electricity needs) is coupled with carbon capture and sequestration (CCS) systems. CCS systems collect carbon dioxide from power plant exhaust gases (using considerable energy in the process) for sequestration in (typically) underground strata.

The SWITCH-China IPCC case results in a reduction of power-sector CO_2 emissions to 80 percent of 1990 levels. The authors of the study report that the annual additional cost of the scenario is over $2 trillion per year by 2050, which is offset at least partially (22 to 42 percent) by the avoided external costs of coal production and use in the IPCC scenario. Placed in context, $2 trillion is about 5-7 percent of projected Chinese GDP in 2050. It should be noted that the SWITCH-China Reference Case appears to call for considerably higher growth in overall generation capacity, and thus likely in generation, relative to the composite BAU scenario described in section 2.2 of this paper.

3.2.4. Commentary on Plans for Nuclear and Coal-fired Power Development by Greenpeace

Although the author of this paper was unable to find long-range projections of the evolution of the Chinese energy sector authored by Greenpeace that are similar to the studies described above, a number of Greenpeace publications have commented on the future of the Chinese coal and nuclear sectors. A 2012 article suggests that

175. Capacity estimates are by the author of this paper based on review of figures in He et al (2016).

China should focus on renewable energy and energy efficiency to reduce greenhouse gas emissions, and that

> "...building enough nuclear power stations to make a meaningful reduction in greenhouse gas emissions would cost trillions of dollars, create tens of thousands of tons of lethal high-level radioactive waste, contribute to further proliferation of nuclear weapons materials, and result in a Chernobyl-scale accident once every decade. Perhaps most significantly, it will squander the resources necessary to implement meaningful climate change solutions."[176]

A late 2017 article notes that China has recently canceled or delayed work on more than 150 planned or under-construction coal-fired power plants due to the "flat-lining" of demand for coal-fired power due to overcapacity of coal-fired generation. The article cites the environmental benefits of not moving forward with coal-fired power, including air pollution and water consumption benefits, and indicates that newly-increased targets for deployment of solar and wind power sources will further reduce the need for coal-fired generation. The article also notes that "[t]he government's recent efforts to clamp down on the red-hot real estate sector and local government debt spending – key drivers of China's heavy industry volumes and power demand – will also leave less space for coal-fired power generation."[177]

---

176. Greenpeace East Asia (2012), "No nuclear in China - not anywhere", dated 10-31-2012, and available as http://www.greenpeace.org/eastasia/news/blog/no-nuclear-in-china-not-anywhere/blog/42752/.
177. Lauri Myllyvirta and Li Danqing (2017), China halts more than 150 coal-fired power plants", dated October 11, 2017, and available as https://unearthed.greenpeace.org/2017/10/11/china-halts-150-coal-fired-power-plants/.

A Greenpeace report evaluated the co-benefits of renewable generation in China from 2015 through 2030.[178] Among other findings, the report:

- Finds that the external environmental benefits of renewable generation exceed the wind power subsidy provided even in 2016[179] and will be more than twice the level of the subsidy by 2030.

- States that "Between 2015 and 2030, wind and solar PV power will contribute RMB 14.3 trillion to China's GDP."

- Estimates that the wind and solar industries will employ 7.7 million people by 2030.

- Indicates that solar energy has been a force in "energy poverty" alleviation by being a major tool for providing electricity to households that previously lacked grid access.

- Notes that wind and solar development has reduced water use by coal-fired power plants, with the reduction by 2030 "... expected to increase to 3.6 billion m^3, equivalent to the annual basic water consumption of 200 million people."

A 2016 article provided by Greenpeace identifies some of the issues affecting deployment of reactors in China based on imported

---

178. Greenpeace (2017), *Accelerating the Energy Transition The Co-benefits of Wind and Solar PV Power in China, Summary* dated April, 2017, and available as https://secured-static.greenpeace.org/eastasia/PageFiles/299371/Renewables%20co-benefits%20report,%20April%2017/Accelerating%20the%20Energy%20Transition_GPEA%20media%20briefing_0411.pdf.
179. The subsidy location cited is n Zhangjiakou, Hebei Province, where the wind power generation subsidy was 0.14 RMB/kWh in 2016.

designs, and suggests that these issues will make China's ambitious nuclear power development targets difficult to achieve.[180]

3.2.5. China National Renewable Energy Center's "China Renewable Energy Outlook 2016" (CREO)

The report *China Renewable Energy Outlook 2016*, prepared by the China Renewable Energy Center in collaboration with groups in China, Europe, and the United States, focuses on an alternative "High Renewable Energy (RE) Penetration" scenario to a "Stated Policies" reference case, both evaluated through 2030.[181] The two scenarios differ only modestly with regard to overall energy consumption, with the Stated Policies case reaching a slightly higher peak in overall energy consumption a few years later than the High RE Penetration case, and having a somewhat slower decline in contribution of coal to primary energy use (47 percent of total primary energy in 2030 in the Stated Policies scenario versus 42 percent in the High RE Penetration scenario, down from 65 percent in 2016). The High RE Penetration case, however, provides a third more RE energy production by 2030 than the Stated Policies case. This difference is accomplished mainly by adding over 1000 GW of mainly wind and solar generation capacity in the High RE Penetration scenario, over and above the amounts assumed for the Stated Policies case. Coal-fired capacity falls from a high of 960 GW in 2020 in both scenarios, to 710 GW in the Stated Policies case

180. Zachary Davies Boren (2016), "All is not well with nuclear in China", dated February 17, 2016, and available as http://energydesk.greenpeace.org/2016/02/17/not-even-china-likes-the-hinkley-epr-nuclear-reactor/.
181. China National Renewable Energy Center and Energy Research Institute (CNREC and ERI, 2016), *China Renewable Energy Outlook 2016*, prepared with cooperation partners the Danish Energy Agency and the US National Renewable Energy Laboratory, and available as http://www.ea-energianalyse.dk/reports/1473_REO2016.pdf.

and 660 GW in the High RE Penetration scenario, Generation from coal-fired power plants falls more dramatically, however, and as a result CO_2 emissions are 1.25 billion tonnes lower (about 13 percent) in the High RE Penetration scenario, and local air pollution and its health effects fall much more rapidly as well. The CREO report estimated a small increase in GDP in the High RE Penetration case relative to the Stated Policies case, with a significant boost to employment in renewable energy-related occupations partially offset by a reduction in employment in the fossil fuels sectors, mostly coal mining.

With regard to the nuclear sector, the CREO report includes the following description:

> "According to the decisions by the previous as well as current government of China, construction of nuclear power plants in the inland and in large-scale construction in the Yangtze River Basin will not happen. The development of the western regions has priority for a "green mountains and clear water are as good as mountains of gold and silver", primarily based on renewable energy. Before the fourth generation of nuclear power technology is in commercial operation, it is assumed that China will not open the inland deployment of nuclear power. Based on this, we consider nuclear power development to be within the range of 100 GW in 2050, solely deployed in coastal areas."

Although nuclear deployment in the two scenarios considered in the CREO 2016 report do not differ markedly, with nuclear development limited to 75 GW by 2030, CREO thus projects much less nuclear capacity than that implied by the roster of planned reactors presented by the World Nuclear Association (as described above), but still creates significant environmental benefits, relative to the Stated Policies case, through the High RE Penetration scenario.

The CREO project partners have been updating and expanded their analyses, and a "CREO 2017" report is forthcoming.

3.2.6. Energy Research Institute's "China 2050 High Renewable Energy Penetration Scenario And Roadmap Study"

The 2015 report *China 2050 High Renewable Energy Penetration Scenario And Roadmap Study*, prepared by the Energy Research Institute (ERI) of the National Development and Reform Commission (NDRC), and funded by the Energy Foundation, presents a scenario for the evolution of the Chinese energy sector that, like the CREO and LBNL studies, includes significant additional electrification, relatively modest growth in nuclear generation capacity and use, and continuing strong growth in the deployment of renewable power sources.[182] Key findings of the report, as relayed in its *Executive Summary*, include the following:

- Total power generation in 2050 is higher than that in the BAU scenario presented in section 2 of this paper, due principally to higher growth in generation in the early years of the projection, though growth in the later years (2040-2050) is somewhat lower in the ERI projection.

- The fraction of generation provided by coal-fired power plants falls from nearly 75 percent in 2011 (the base year of the ERI study) to less than 7 percent in 2050.

---

182. Energy Research Institute (2015), *Executive Summary, China 2050 High Renewable Energy Penetration Scenario And Roadmap Study*, dated April, 2015, and available as http://www.efchina.org/Attachments/Report/report-20150420/China-2050-High-Renewable-Energy-Penetration-Scenario-and-Roadmap-Study-Executive-Summary.pdf.

- The fraction of generation provided by solar and wind power rises to nearly 54 percent of the national total by 2050, with hydropower providing another 14 percent.

- Nuclear power generation capacity grows to 100 GW by 2050, at which point it supplies 4.3 percent of total generation. In the ERI projection, nuclear power's share of generation stays between 4.8 percent and 3.7 percent from 2018 through 2050.

- Carbon dioxide emissions fall to about 3 GT CO2 by 2050.

- Emissions of the local and regional air pollutants nitrogen oxides and sulfur oxides (NOx and SOx) fall to approximately 1970 levels (NOx) and about half of 1970 levels (SOx), on the order of a tenth of peak emissions of both gases (reached in 2010 and 2005, respectively).

- The average costs of generating and delivering electricity in ERI's high renewables scenario are only very modestly higher than in ERI's reference case, ranging from 0.672 RMB/kWh in 2030 to 0.685 RMB/kWh (about 10.4 U.S. cents/kWh at current exchange rates) in 2050 in the high renewables case, "while in [the] reference scenario the average cost between 2030-2050 will stay flat around RMB 0.67/kWh."[183]

- Under the high renewables scenario, about 10 million jobs are added in the renewable energy industries between 2015 and 2050.

---

183. Quote from ERI (2015), ibid. Note that projecting costs two or three decades into the future is necessarily an uncertain exercise, thus the main point here is that the costs in the high renewables case appear minimally different from those in the reference case.

- Relative to the reference case, ERI's macroeconomic analysis of the high renewables scenario indicates modest changes in key economic indicators in 2050. Government spending is lower, imports and exports higher, and residential consumption up, but all within the range of a 0.23 and 1.44 percent difference between cases. Price levels in the two cases are virtually the same in 2050.

Overall, ERI's results suggest that a shift to a high renewables penetration scenario can provide very significant environmental benefits at minimal overall costs to the economy—though of course some sectors will come out better than others—and without large growth in the deployment of nuclear power.

3.3. Quantitative and Qualitative Comparison of Alternative Scenarios with Baseline

The alternative scenarios of and commentary on China's near- and medium-term energy future presented above indicated that serious consideration is being given both inside and outside of China to policies to transform China's economy and energy sector towards a much lower-carbon, higher-renewables, and more efficient system than exists today. Some of the works described above (which certainly does not exhaust the universe of China energy futures studies) posit a relatively modest role for nuclear power in China's electricity sector by 2050, with little or no growth over time in nuclear's share of generation, while others suggest a more important role. In reviewing all of the projections above, and those presented in this paper, it is important to consider the breathtaking changes ongoing in the Chinese energy sector. Trends in just the past few years—marked reductions in deployment and use of coal-fired power, marked reductions in the rate of growth of electricity demand, and vast increases in wind and solar generation capacity (along with reductions in wind and solar costs) — are shifting the baseline upon which scenarios are constructed. These trends may rapidly make readily achievable scenarios that look unachievable today.

Below we present, in the broader context of the overall electricity-sector scenarios described above and in section 2 of this paper, three alternative scenarios of nuclear power sector development in China, and offer a brief analysis of their relative benefits and costs, both quantitative and qualitative.[184]

3.3.1. Summary of Nuclear Capacity Scenarios

In order to explore the consequences of different scenarios of nuclear power capacity development in China, we have created three different capacity expansion cases. These are as follows:

- A "**Business as Usual**", or **BAU** case that draws from a very recent listing of planned and proposed reactors in China prepared by the World Nuclear Association (WNA),[185] but assumes that the phase-in of new "planned" reactors will be somewhat slower than in the WNA listing,[186] and that about

184. The analysis of benefits and costs defines "energy security" in a broad sense, including traditional energy supply security and net direct costs, but also including environmental, social and political, and military-related security. For a discussion of an analytical framework used with this broader concept of energy security, see for example, David von Hippel, Tatsujiro Suzuki, James H. Williams, Timothy Savage, and Peter Hayes (2011), Energy Security and Sustainability in Northeast Asia, published in the "Asian Energy Security" Special Section of *Energy Policy* Volume 39, Issue 11, November, 2011, pages 6719–6730.
185. World Nuclear Association (2017), "Nuclear Power in China", updated July 2017, available as http://www.world-nuclear.org/information-library/country-profiles/countries-a-f/china-nuclear-power.aspx.
186. Some authors see evidence that China's nuclear build-out is not proceeding as rapidly as initially planned. See, for example, Steve Thomas (2016), "China's

55 GWe of the reactors listed as "proposed" by WNA will ultimately be built by 2050 (out of a total of 200 GWe of projects listed). Based on standard operating lifetimes, existing LWRs are retired when they reach 40 years of service, and China's two "CANDU" reactor units are retired after 30 years of service. By 2050, nuclear generation capacity in the BAU case reaches a level that is slightly lower than the nuclear capacity projected in the USDOE EIA's *International Energy Outlook 2017* reference case, and also slightly lower than estimated 2050 nuclear capacity the *Reinventing Fire* China reference case.

- A "**Maximum Nuclear**" or **MAX** case, which also draws from the WNA listing but assumes a faster phase-in than in the BAU case of under-construction and "planned" units, and also assumes that about 80 percent (160 GWe) of the reactors listed in the World Nuclear Association table as "proposed" (or replacements similar in total capacity) are ultimately built and are all phased in by 2050. Existing LWRs are retired when they reach 50 years of service, so 10 years of life extension is assumed. By 2050, nuclear generation capacity in the MAX case is lower than that implied in RMI's Reinventing Fire scenario.

- A "**Minimum Nuclear**" or **MIN** case, in which the reactors listed by the World Nuclear Association as under construction or planned are all ultimately phased in by 2038, though on a much slower schedule (for those plants for which an operational date is given) than listed. After 2038, no additional plants are built, and as in the BAU case, older plants are retired as they reach the end of their standard

Nuclear Power Plans Melting Down", *The Diplomat*, October 29, 2016, available as http://thediplomat.com/2016/10/chinas-nuclear-power-plans-melting-down/.

operating lives. This lower capacity scenario could arise due to a combination of factors, including minimal growth in electricity demand in the 2030s and beyond, increased price competition from renewable energy, the availability of new and cost-effective electricity storage technologies, and/or perhaps a social backlash against nuclear power, In the MIN case, nuclear capacity begins to slowly decline around 2050, reaching a level by 2050 that is somewhat below that of the ERI and CREO scenarios described in section 3.2.

Readers should note that none of these paths account explicitly for potential shocks to the Chinese nuclear power industry, and to Chinese society as a whole, that might arise from a serious accident in a Chinese nuclear power plant. Such an event could have potentially devastating consequences for large populations.[187] The timing of such an event, should it occur, is not knowable in advance, although there is an argument that it is statistically likely over the time frame of these paths, given historical rates of major accidents per year of reactor operation.[188]

Figure 0-1 and

---

187. The author of this paper prepared a rough estimate of the consequences of an incident involving (accident at or attack on) a reactor and spent fuel pool at the Ling'Ao nuclear plant in Guangdong province that suggested that "worst case" releases of Cesium-137 might result in exposures sufficient to cause hundreds of thousands of premature cancer deaths and almost certainly require the abandonment of one or several big cities, depending on the prevailing wind direction at the time of the incident (D. von Hippel and P. Hayes, 2016, unpublished).

188. See, for example, He Zuoxiu, "Chinese nuclear disaster "highly probable" by 2030," *China Dialogue,* March 19, 2013, at: https://www.chinadialogue.net/article/show/single/en/5808, and S. Wheatley, B. Sovacool, D. Sornette (2017), "Of Disasters and Dragon Kings: A Statistical Analysis of Nuclear Power Incidents and Accidents," *Risk Analysis*, Volume 37, #1, pages 99-115, available as http://onlinelibrary.wiley.com/doi/10.1111/risa.12587/epdf.

Figure 0-2, respectively, show the capacity and electricity output implied by each of the three nuclear scenarios above. In the BAU case, capacity increases to 170 GWe by 2050, while in the MAX case capacity rises to nearly 260 GWe. In the MIN case, capacity rises to about 91 GWe by 2040 (about 260% of existing capacity as of 2017), but no new plants are added thereafter, so capacity falls to about 85 GWe by 2050 as older plants are retired. By way of comparison, as noted above, China's current (2016) overall electricity generation capacity for all types was over 1600 GW, and overall generation was nearly 6200 TWh.[189]

Figure 0-1: Three Scenarios of Nuclear Generation Capacity (GWe) for China through 2050

189. For electrical energy, see British Petroleum (2017), "BP Statistical Review of World Energy June 2017" workbook, available as http://www.bp.com/content/dam/bp/en/corporate/excel/energy-economics/statistical-review-2017/bp-statistical-review-of-world-energy-2017-underpinning-data.xlsx. For capacity, see, for example, "Installed capacity of electric power generation in China between 2010 and 2016 (in GW)", available as https://www.statista.com/statistics/302269/china-installed-power-generation-capacity/.

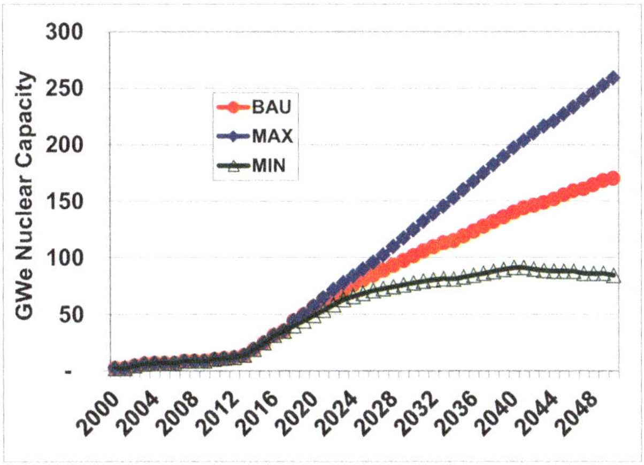

Figure 0-2: Three Scenarios of Nuclear Generation Output (TWhe) for China through 2050

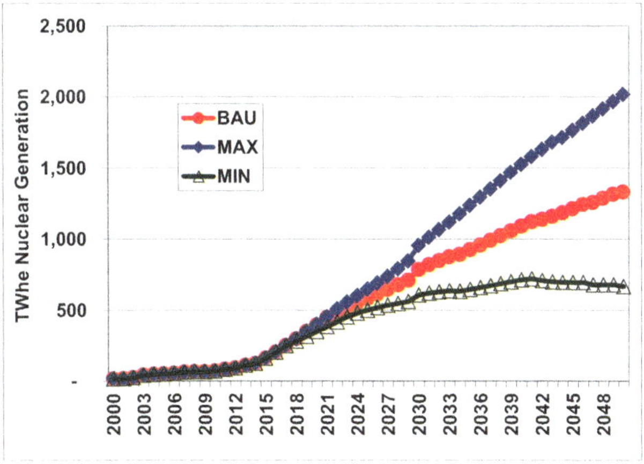

3.3.2. Broader Energy and Nuclear Sectors Contexts for Nuclear Paths

Each of the nuclear generation capacity expansion paths could, in theory, be combined with a range of different trends and policies in the broader energy and electricity sectors, as well as in the overall nuclear sector, including front-end and back-end fuel cycle developments. In practice, competition for private and public investment resources, national and societal goals and preferences, and other criteria make some combinations more plausible than others. The assumed energy and nuclear/sector contexts for each of the paths described above is as follows.

BAU Path

The BAU path is assumed to exist within a Chinese energy sector in which significant, but not aggressive, efforts to displace coal-fired power with renewable energy sources continue, energy efficiency improvements also continue but are not major priorities of national energy policy, and some electrification of currently fossil-fuel-dominated sectors and end-uses, most notably transport, occurs, but is again aggressively pursued. In the BAU path, investment priorities in the energy sector are thus split between the nuclear and renewables sectors, with some development of coal- and gas-fired power continuing.

In the BAU path, as well as in the MAX and MIN paths as described below, China is assumed to source one-third of its uranium domestically.[190] In the nuclear sector under the BAU path, the

190. From World Nuclear Association (2017), "China's Nuclear Fuel Cycle, updated September, 2017, and available as http://www.world-

development of uranium enrichment and reprocessing facilities, and the use of MOx fuel, proceeds roughly as described by the World Nuclear Association, but on a somewhat delayed schedule. For enrichment, this means that about 80 percent of China's enrichment needs are supplied domestically by 2020, with enrichment being entirely domestic by 2030. World (and Chinese) enrichment costs, driven in large part by Chinese nuclear expansion, follow a "medium" scenario, rising to about $75/kg SWU (about twice 2017 levels, but much lower than historical maxima) by 2050 (2009 dollars). Research and pilot development of fast reactor technologies continues, but also on a somewhat delayed schedule, such that commercialization of fast reactors is still at least several years off by 2050. MOx use in LWRs is assumed to start in 2025,[191] and 25 percent of reactors are assumed to use cores with 20% MOx fuel by 2050. Reprocessing also starts in 2025, with capacity and throughput ramping up sufficiently to process 50 percent of cooled spent fuel by 2040 and thereafter.

Spent fuel management in the BAU path, consistent with the path's emphasis on reprocessing, focuses mainly on interim spent fuel storage in spent fuel pools at reactors and at reprocessing facilities, although there is some dry cask storage of spent fuel.[192]

---

nuclear.org/information-library/country-profiles/countries-a-f/china-nuclear-fuel-cycle.aspx. "China aims to produce one-third of its uranium domestically, obtain one-third through foreign equity in mines and joint ventures overseas, and to purchase one-third on the open market."

191. See, for example, Qiang Yue, Jingke He, Laurence Stamford, and Adisa Azapagic (2017) "Nuclear Power in China: An Analysis of the Current and Near-Future Uranium Flows", *Energy Technologies*, 2017, Volume 5, pp. 681 – 691, available as http://onlinelibrary.wiley.com/doi/10.1002/ente.201600444/epdf

192. "Dry casks" refers to the spent fuel storage option, used in many countries, in which cooled spent fuel is isolated in massive casks designed to last for up to 100 years. Spent fuel assemblies are typically sealed into stainless steel containers several centimeters thick, which are then placed in "overpacks" made of steel or concrete. The resulting cask is on the order of 2 meters in diameter and 6 meters tall, weighs up to 50 tonnes, and is essentially impervious to any

MAX Path

Consistent with, for example, the Reinventing Fire scenario described above, the MAX Path includes both significant investments in energy efficiency and in renewable generation, as well as in electricity transmission facilities. Electrification of the energy sector is also a priority, with the reduction of greenhouse gas emissions and local/regional air pollutants a driving policy impetus. Both the savings in year 2050 electricity needs due to energy efficiency improvements and additional electrification to displace fossil fuel at the end-use level are assumed to be about half that assumed in the Reinventing Fire study—about 25 and 17 percent, respectively. The lowered assumptions for these impacts are in part due to a lower reference case assumed here than assumed in the Reinventing Fire study, and in part to an assumption that a more aggressive build-out of the nuclear sector, and of, for example, reprocessing (see below) and enrichment, will to some extent crowd out investments in efficiency and renewable energy (for example). Investment to drive the nuclear sector comes to a large extent from public funding, while renewable power investments are mostly privately driven but take advantage of necessarily higher prices for electricity required by nuclear investments.

The MAX path focuses on building national capacity in the nuclear sector, including aggressive build-outs of uranium enrichment and related capacity, as well as reprocessing capacity and MOx use. For enrichment, this means that about 90 percent of China's enrichment needs are supplied domestically by 2020, with enrichment being entirely domestic by 2025. Enrichment costs, again driven by Chinese demand, are assumed to follow a "high" scenario, rising to

---

significant damage that could be caused by natural disasters or accidents, as well as to all but the most determined and well-equipped attempts at penetration by criminals or terrorists.

$104/kg SWU (still much lower than the $160/kg SWU historical maxima) by 2050. National capacity for manufacturing of key nuclear plant components grows rapidly, and the MAX path is likely to be also combined with an aggressive national effort to export nuclear reactors to other nations. MOx use in LWRs is assumed to start in 2022, and 40 percent of reactors are assumed to use cores with 20% MOx fuel by 2050. Reprocessing on a commercial scale likewise starts in 2022, with a target of 80 percent of cooled spent fuel reprocessed by 2035. Spent fuel management focuses on interim storage in spent fuel pools at reactors and at reprocessing facilities.

MIN Path

Consistent with paths projecting high rates of renewable energy penetration in the CREO and ERI studies, the MIN path couples nuclear capacity expansion at a rate that holds the nuclear share of generation roughly constant with aggressive development of renewable energy for power generation and for end-uses, plus aggressive energy efficiency programs. Electrification is also an emphasis, both for greenhouse gas emissions reduction and for bringing down emissions of local and regional air pollutants. Energy efficiency efforts, and the use of more renewable energy sources at the end-use level, are assumed to reduce electricity needs by 33 percent by 2030 relative to BAU requirements, with electrification adding 20 percent back to electricity demand by 2050. Ongoing policies include carbon markets and carbon taxes, preferentially target markets for renewable energy and energy efficiency, with the proceeds used to support investment in both, as well as in pollution control and environmental remediation. The MIN path does not explicitly include carbon capture and sequestration (CCS) for coal- and/or gas-fired power plants, as used in the "IPCC Target" scenario of the Gang He, et al (2016) study referenced above, but could include CCS if CCS technology is suitably advanced and if greenhouse gas emissions reduction policies in China required its use. China has been rapidly developing facilities for importing LNG, and LNG could play a more important role in power generation in China's future in a variant of a MIN path, with or without CCS. Greater gas-fired generation, perhaps displacing some planned nuclear plants, would be even more plausible if, for example, infrastructure for bringing North American gas to China at attractive prices can be developed. Also not explicitly modeled in the MIN path is the development and widespread use of electricity storage technologies, which would be needed to complement the aggressive development of renewable (wind and solar) electricity sources.

In the nuclear sector, reprocessing is not pursued beyond the existing and under-construction pilot plants, and MOx use in existing reactors is limited. Uranium enrichment facilities planned for the near-term are built by 2025, but no additional enrichment plants are built, and the remainder of China's required enrichment services are imported. With a smaller nuclear capacity expansion in China, and (assumed) reduced nuclear restarts in Japan, relative to other scenarios, international (and Chinese) enrichment costs follow a "low" trajectory, reaching $54/kg SWU by 2050, about 40 percent above 2017 levels. Efforts to export reactors continue, but are not heavily subsidized by the Chinese government. Research into fast reactors continues, but at a low level, and commercialization of fast reactors is still decades away by 2050. In part to provide fuel for fast reactor research, the smaller (200 tHM/yr) reprocessing facility reportedly (as of 2017) being built at Gansu is eventually completed, but is not run until 2025, and then at only partial capacity for several years. Pu from this facility is used to produce MOx fuel, which is used in 10 percent of LWRs by 2050, with phase-in occurring slowly starting in 2030. MOx fuel again makes up 20 percent of reactor cores in those units that use MOx. Cooled spent fuel is stored mostly in a mixture of at-reactor spent fuel pools and in dry casks, which may be located at or near reactors or at a centralized dry cask facility.

3.3.3. Flows of Nuclear Materials

Table 0-1 shows estimated total requirements for uranium and uranium ore in each of the three scenarios, both for the individual years 2010, 2030, and 2050, and on a cumulative basis from 2015 through 2050. The MAX path implies the use of over 50,000 tonnes of U annually by 2050, while less than a third as much is required in the MIN path. The extraction of 11 million tonnes of ore is required by 2050 in the MAX path, but placed in context, this is much less than a percent of the total volume of coal extracted to fuel China's power sector. By assumption, two thirds of Chinese uranium needs are sourced abroad, and one third are from domestic mines.

Table 0-1: Uranium Requirements by Nuclear Capacity Expansion Path

Parameter	YEAR	MAX Path	BAU Path	MIN Path
Annual Total Metric Tons Natural Uranium (as U) Imported plus Domestic Production	2010	2,629	2,629	2,629
	2030	27,718	21,837	15,946
	2050	53,659	35,106	16,570
	Cumulative, 2015-2050	1,105,629	820,318	540,581
Annual Total Thousand Metric Tons Uranium Ore (from In-country and outside mines) to Supply All Domestic Uranium Needs	2010	546	546	546
	2030	5,752	4,532	3,309
	2050	11,136	7,286	3,439
	Cumulative, 2015-2050	229,457	170,245	112,190

Under the MAX expansion path, if China chose to provide all of its own enriched uranium, China alone would need to build new enrichment capacity by 2050 approximately equal to more than half of today's global capacity. China's annual requirement requirements by 2050 rise to nearly 42 Million kg SWU. Under the MIN expansion path, however, international enrichment facilities extant as of 2015 are likely sufficient to meet China's enrichment needs by 2050 (about 12 million kg SWU), even factoring in likely East Asia regional and out-of-region demand without significant expansion, assuming existing international enrichment facilities (or replacement facilities) continue to operate. Though the ROK and Japan have accounted for almost all enriched uranium in East Asia needs pre-Fukushima, the rapid growth of China's nuclear power sector and the slow process of restarting Japan's reactors means that China's demand for enrichment will likely outstrip needs in the rest of the region well before 2020.

The uranium oxide (UOx) and MOx fuel requirements under each scenario are summarized in Table 0-2. MOx fuel requirements in the BAU path by 2050 are less than half of those in the MAX path, and MIN path MOx use is less than a tenth of MAX path use by 2050.

Table 0-2: Requirements for UOx and MOx Fuel in Each of 3 Nuclear Paths

Parameter	YEAR	MAX Path	BAU Path	MIN Path
Implied Total Requirements for UOx Fuel (excluding MOx) from All Sources (Metric tonnes heavy metal in fabricated fuel)	2010	267	267	267
	2030	2,929	2,327	1,711
	2050	5,417	3,620	1,763
	Cumulative, 2015-2050	115,040	86,601	57,961
Implied Requirements for MOx Fuel Blended and Fabricated from All Sources for Use in Domestic Reactors (Metric tonnes heavy metal in fabricated fuel)	2010	-	-	-
	2030	60	24	-
	2050	408	191	36
	Cumulative, 2015-2050	4,657	2,123	394

With widely varying use of reprocessing in the three scenarios, it is not surprising that plutonium production and uptake (as MOx) also is significantly different, as shown in Figure 0-3. Cumulative plutonium separation through reprocessing rises to over 450 tonnes of Pu in the MAX case by 2050, and about 9 tonnes of Pu inventory remains after MOx use in 2050. About 200 tonnes of Pu are produced by 2050 in the BAU case, but almost all stocks are used as MOx. In the MIN case, only about 27 tonnes of Pu are produced via reprocessing in China through 2050, and more than that amount of Pu is used as MOx, implying, for example, that some Pu from other nations (stocks from Russia, the UK, or France) is blended into fuel used in Chinese reactors, and/or military Pu from the Chinese weapons program is disposed of as MOx. Note that all of these calculations of net Pu inventories by 2050 are extremely sensitive to the combination of assumptions regarding the fraction of spent fuel reprocessed and the amount of fuel used as MOx. In practical terms, this means that if a reprocessing program is successful BUT MOx use is delayed, significant inventories of Pu can build up, and can serve as a proliferation target. As a sensitivity analysis, Figure 3-4 shows what might happen if reprocessing proceeds in each path as indicated above, but MOx use is delayed by 10 years in each path, and used in only half as many reactors—a plausible outcome given the difficulties in implementing MOx use that other nations have experienced to date. In this sensitivity case, inventories of Pu of about 300, 140, and 18 tonnes build up in the MAX, BAU, and MIN cases, respectively, by 2050, at which point Pu stocks are still continuing to accrue. The stocks building up in this sensitivity case in the MAX and BAU paths would suffice to build tens of thousands of nuclear weapons, and even the MIN case stocks represent thousands of times the mass of Pu contained in a nuclear warhead. Additionally, paths that produce high volumes of Pu, whether or not the Pu is fully consumed as MOx, would tend to enhance the chance of significant volumes of Pu going astray, as the more Pu is produced, the easier it will be for weapons-relevant quantities (kilograms) of Pu, amounting to less than a tenth of a percent of MAX and BAU annual output from reprocessing facilities, to be diverted for criminal purposes.

Figure 0-3: Pu Separation and Stocks Net of MOx Fuel Use by Path

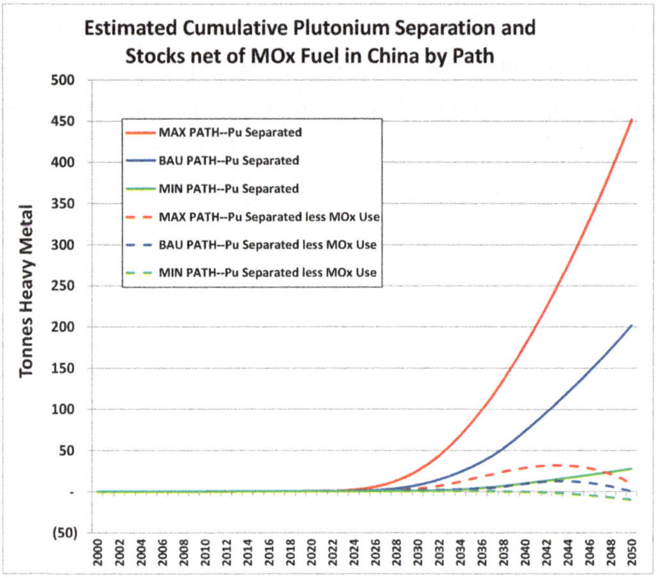

Figure 0-4: Sensitivity Analysis: Pu Separation and Stocks Net of MOx Fuel Use by Path if MOx Use is Delayed

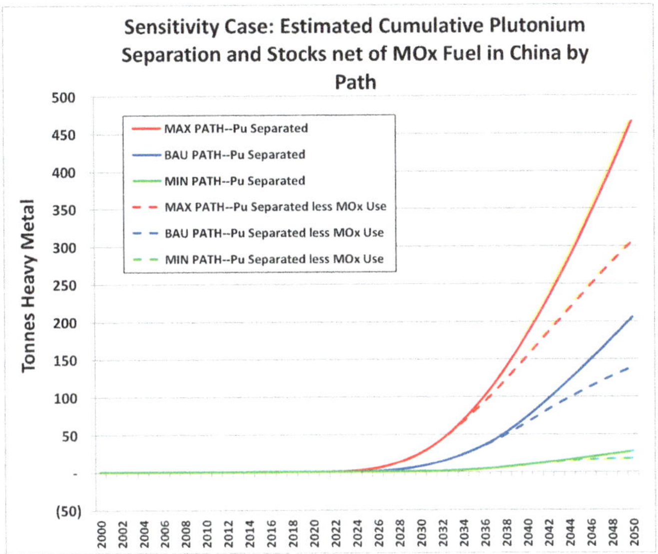

Table 0-3 presents annual and cumulative results for the production of cooled spent fuel[193] and high-level wastes from reprocessing, as well as the number of casks used for dry cask storage, in each of the three nuclear paths. Despite the large difference in 2050 nuclear generation capacity between the three paths, the cooled spent fuel produced over time is not all that different, because cooled spent fuel production lags changes in generation. The MAX scenario produces much more high-level waste (HLW) from reprocessing than the other cases. Though the volume of HLW is not large—5000 or so cubic meters could be contained in 25 or so average urban apartments—it is highly radioactive and, like spent fuel, remains so

193. Spent LWR nuclear fuel must be cooled—typically for five to 10 years, in spent fuel pools before it can be either reprocessed, placed in dry casks for long-term (100 years or more) storage, or placed in long term storage/indefinite or permanent disposal in mined repositories, such as the canceled Yucca Mountain facility in the US, or in deep borehole disposal (a technology in the early phases of development—see, for example, Neil Chapman (2013), "Deep Borehole Disposal of Spent Fuel and Other Radioactive Wastes", NAPSNet Special Reports, July 25, 2013, https://nautilus.org/napsnet/napsnet-special-reports/deep-borehole-disposal-of-spent-fuel-and-other-radioactive-wastes/.

for thousands of years, meaning that special well-secured facilities capable of holding the wastes indefinitely must be constructed. In the MIN path, where the emphasis is on dry cask storage, on the order of 2600 casks would be required to accommodate the cooled spent fuel produced by 2050, not including the relatively small amount of spent fuel reprocessed in the MIN path. To put this number of casks in perspective, the total dry casks filled during the MIN path could be stored in an area of less than 10 hectares, just a bit bigger than area enclosed by the fence around the White House in Washington DC.

Table 0-3: Spent Fuel Management Results for Three Nuclear Paths

Parameter	YEAR	MAX Path	BAU Path	MIN Path
Annual New Spent LWR Fuel Cooled and Available for Reprocessing, Storage, or Disposal (excluding MOx spent fuel), Metric Tonnes Heavy Metal	2010	51	51	51
	2030	1,190	1,139	975
	2050	3,654	2,686	1,747
	Cumulative, 2015-2050	57,028	46,688	35,880
Annual Spent MOx Fuel Cooled and Available for Storage or Disposal, Metric Tonnes Heavy Metal	2010	-	-	-
	2030	-	-	-
	2050	239	112	23
	Cumulative, 2015-2050	2,020	883	150
Implied Amount of Spent LWR Fuel (UOx only) Reprocessed in Total for Use in Domestic Reactors (Metric tonnes heavy metal)	2010	1	1	1
	2030	612	214	9
	2050	2,923	1,343	175
	Cumulative, 2015-2050	41,045	18,276	2,438
Implied Volume of High-level Waste (as vitrified) from All Cooled Spent LWR Fuel (UOx only) Reprocessed for Use in Domestic Reactors (cubic meters)	2010	0.15	0.15	0.15
	2030	70.37	24.57	1.02
	2050	336.18	154.47	20.10
	Cumulative, 2015-2050	4,720.15	2,101.75	280.35
Total Annual Number of Casks Required for Dry Cask Storage of Cooled Spent LWR Fuel (UOx only) for Storage/Disposal, Net of Reprocessing (units)	2010	-	-	-
	2030	-	-	-
	2050	36	109	156
	Cumulative, 2015-2050	362	1,496	2,608
Total Annual Number of Casks Required for Dry Cask Storage of Cooled Spent LWR MOx Fuel for Storage/Disposal (units)	2010	-	-	-
	2030	-	-	-
	2050	24	11	2
	Cumulative, 2015-2050	202	88	15

3.3.4. Costs

Although a full quantitative evaluation of the direct costs of the scenarios described above is beyond the scope of this paper, some of the considerations in comparing the costs between the three scenarios in qualitative terms are as follows:

- Fossil fuel costs for electricity generation will be highest for the BAU case, with annual costs in the MAX case about half of BAU levels by 2050, and in the MIN case about a quarter of BAU levels, due to displacement of fossil fuels by efficiency improvements, electrification, and use of renewable energy for direct end-uses and electricity generation.

- Overall fossil fuel costs, including "upstream" costs for oil refining, coal mining, and other fuel-cycle activities, will be lower due to electrification of end-uses in the MAX scenario, and even lower in the MIN case.

- There is likely to be very little growth in capital and operating and maintenance (O&M) costs for fossil-fueled power plants in any of the scenarios, though these costs will be lower in the MAX path, and lower still in the MIN path, as more coal-fired generation is displaced by nuclear (in the MAX path) and renewable generation/energy efficiency in both paths. In the MAX and especially the MIN path, at least more coal-fired capacity will likely be retired, providing a reduction in O&M costs. One factor that could cause power plant costs to rise, and could be applied to any of the three scenarios, is if more stringent controls are required to reduce local and regional air pollutants, and/or if carbon capture and sequestration is required on fossil-fueled plants.

- Costs for energy efficiency and increased deployment of end-use renewables will be higher in the MAX case than in the BAU case, and higher still in the MIN case, although previous experience in China and elsewhere suggests that in many cases energy efficiency provides electricity savings more cheaply than electricity case be generated by supply-side resources. Thus when the avoided costs of electricity generation and direct fuel use are factored in, efficiency investments will likely yield a net savings in overall direct costs. The much higher rate of renewables deployment in the MIN path will likely reduce the per-unit costs of renewable energy systems, and accompanying electricity storage systems. It is difficult to say whether the net costs of aggressive deployment of energy efficiency and renewable energy in the MIN path will be higher or lower than the costs of the conventional and nuclear energy systems that they displace in the BAU and MAX paths, but recent experience and at least some previous studies of energy futures suggest that the cost differences may be small relative to overall energy system costs and to the uncertainties of costs projected three decades into the future.

- The MAX scenario will have the highest overall costs for nuclear generation capacity, as well as for fixed and non-fuel variable operating and maintenance (O&M) costs, followed by the BAU case. MIN case total capacity costs and O&M will be substantially lower than in the other two cases. Costs for fast reactor research and development will be highest in the MAX case, and relatively limited in the MIN case.

Nuclear fuel cycle costs—exclusive of reactor capital and non-fuel O&M costs—have been quantified for each of the three nuclear paths described above. Not surprisingly, as shown in Figure 0-5, nuclear fuel cycle costs are much higher in the MAX path, at a about

$560 billion on a cumulative basis over 2015 through 2050. This total is nearly twice as much as for the BAU path, and on the order of five times that of the MIN path. The three largest cost categories are raw uranium and uranium enrichment—higher in the MAX path due to higher cost escalation assumptions—and reprocessing costs, which are nearly avoided altogether in the MIN path.

Figure 0-5:

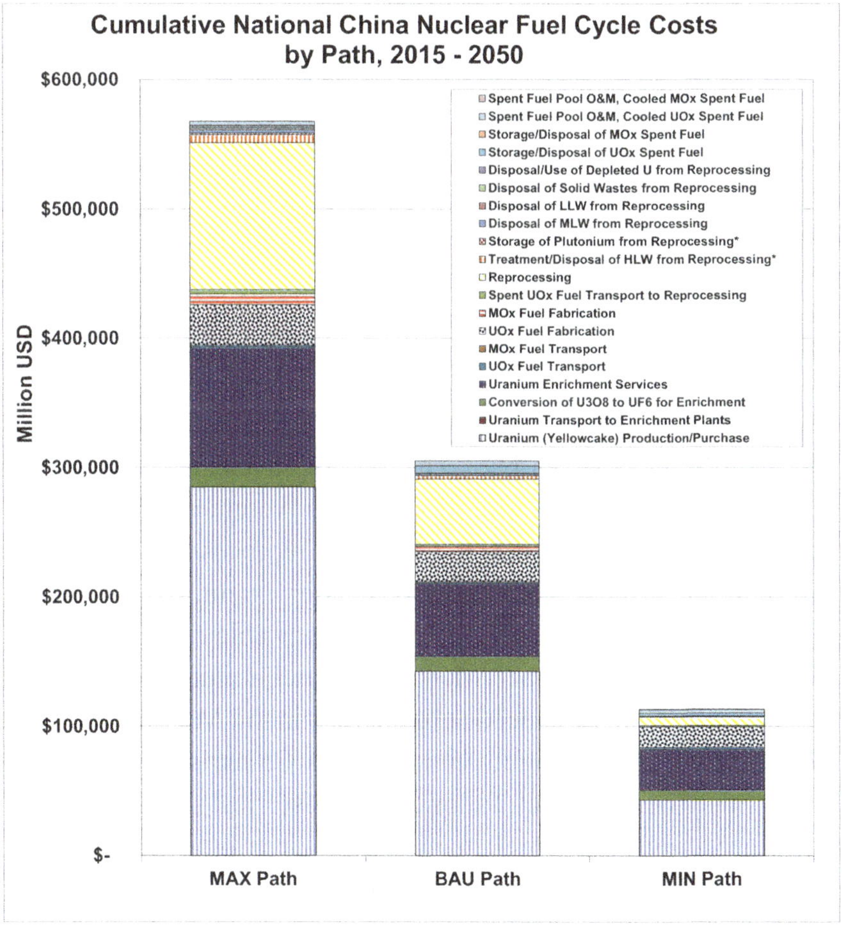

The relative indirect costs to the economy of alternative energy scenarios, and the policies that will drive them, are often a key consideration for policymakers. The perceived and projected impacts of different nuclear scenarios on, for example, GDP and employment at the provincial and national levels in China will have a considerable impact on the acceptability of particular scenarios and policies. In practice, there are always winners and losers—for example, with regard to employment in different sectors and even regions—when policies are shifted, but the net impact of these changes is very hard to know in advance, due to policy- and non-policy-related shifts in the economy and technology (and in underlying costs and factor prices) shifts over time, and to the general uncertainty associated with any economic prognostication. In many macroeconomic studies of different energy scenarios, changes to GDP and employment by, for example, 2050 come out looking like large numbers, but are invariably swamped by the underlying combination of the size of overall GDP by a target year, and the uncertainties inherent in the analysis.

That said, there will be certain sectors that will doubtless win and lose to different extents between the three nuclear scenarios. The nuclear sector, and particularly firms and government organizations associated with advanced fuel cycles and reprocessing, will be the losers in the MIN scenario, and to a lesser extent, the BAU case, relative to the MAX scenario, while the renewable energy industries will benefit most in the MIN case. Coal mining income and employment will be reduced substantially in the MIN case relative to the other cases, though coal mining in China is becoming less labor-intensive in general, following the historical trend in the U.S. and other places. In general, many studies have found that scenarios that focus on renewable energy and energy efficiency produce more long-term jobs and similar if not greater national and regional net income than scenarios focused on supplies of conventional and nuclear energy, but, as noted, there will inevitably be subsectors and industries that are winners and losers in any case relative to any other.

3.3.5. Energy Supply Security

The traditional concept of energy supply security, in brief, is that the more a country can source its fuel requirements from its own territory—or failing that, the more a country can draw for its energy needs from a diversity of domestic resources and imports from large number of trading partners—the more energy supply security is enhanced. Under that definition, the MAX case, which uses more nuclear power and less oil (the reduction being mostly due to electrification of the transport and other sectors), coal, and natural gas, is provides arguably more energy security than the BAU case. The MIN case, with more use of domestic resources (wind, solar, and energy efficiency) than the other two cases, arguably comes out on top in terms of energy supply security. Figure 0-6 shows the relative use of fossil fuels, both cumulative and for selected years, under each case. By assumption, in each case, uranium is sourced from the same ratio of domestic and foreign sources, though in principle, a higher proportion of domestic U could be used in the BAU and, especially, the MIN cases, given the lower overall U requirements.

Figure 0-6: Annual Fossil Fuel Use for Electricity Generation by Type and Year for Three Paths

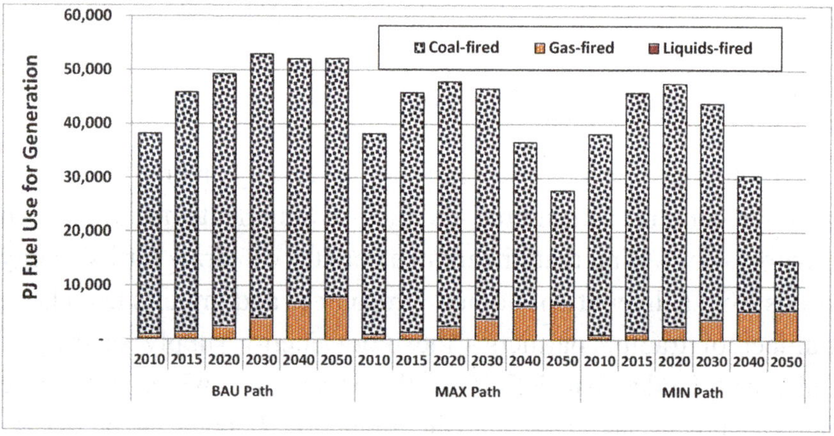

3.3.6. Environment

The environmental component of a broader concept of "energy security" includes comparing the performance of the three scenarios on the basis of emissions of greenhouse gas emissions, local and regional air pollutant production, water pollution, and disposal of nuclear-fuel-cycle related wastes. With respect to these criteria, the three nuclear paths described above yield the following quantitative and qualitative comparisons:

- The MIN path produces nearly 30 percent lower cumulative electricity generation sector (2015-2050) greenhouse gas emissions than the other BAU path, and in particular, as shown in Figure 0-7, produces a small fraction of the year-2050 emissions included in the BAU and MAX paths. Although the overall Chinese societal GHG emissions are the main concern here, and have not been quantified for this paper, the inclusion of additional electrification in the MAX path indicates that China's MAX-path GHG emissions will be considerably lower than in the BAU path, and emissions in the MIN path will be lower still.

- Likewise, though not directly quantified here, non-GHG air pollutant emissions of consequence to local and regional air quality will be less in the MAX path than in the BAU Path, with emissions in the MIN path considerably less in the other two cases, particularly by 2050.

Figure 0-7: Greenhouse Gas Emissions from the Electricity Generation Sector in China Under Three Nuclear Power Development Paths

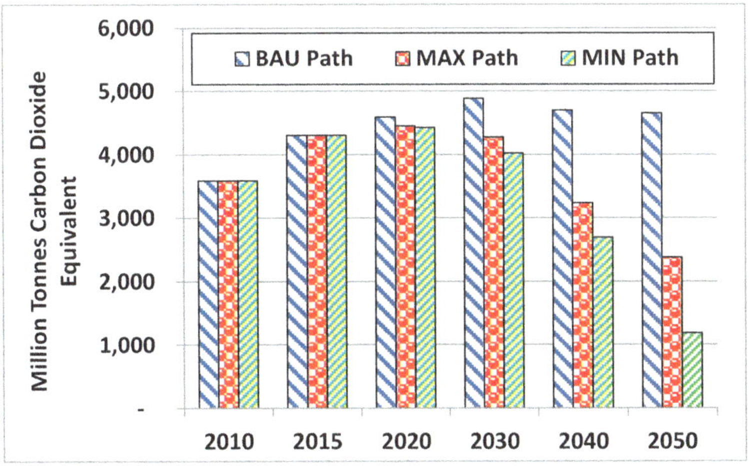

- Because considerably less coal is used in the MIN and MAX paths than in the BAU path, water pollution from coal mines will be less in those paths. Thermal pollution from coal-fired power plants will also be less, though additional thermal pollution from nuclear power plants, some of which may be on inland sites (on rivers), may occur in the MAX path, relative to the other two paths. Additional water pollution and solid wastes from uranium mining in China and abroad will accrue in the BAU and, especially, MAX paths relative to the MIN path.

- The disposal of nuclear-sector-related waste streams will be much more of an issue in the BAU and, especially, MAX paths relative to the MIN path. Considerable HLW and intermediate and low-level wastes from reprocessing will accrue in the BAU and MAX paths, and China will need to find a final resting place for those materials, as well as for any spent fuel placed in long-term storage/disposal. Finding places to store/dispose of these materials may prove to be foci for political and social problems, as noted below.

3.3.7. Social and Political Criteria, and Military Security

During China's period of rapid economic growth, Chinese decisionmakers have typically given (or at least, exhibited) limited concern to the reaction of local populations in decisions on siting of key energy-sector facilities, relative to decisionmakers in many Western nations. Over the past decade, however, trends have suggested that the role of Chinese civil society in the siting of large and potentially polluting or dangerous facilities has been growing, at least in some ways. Very recent events have arguably suggested that civil society's voice in China may not continue to develop as some in the West might have hoped, though the impact of recent changes in Chinese governance on the nuclear sector is not yet clear.[194] In general, paths, like the MAX path, that call for large, centralized, secure facilities for handling and managing nuclear materials may galvanize opposition to such facilities on the local and

194. See, for example, Chris Buckley and Adam Wu (2018), "Ending Term Limits for China's Xi Is a Big Deal. Here's Why", *The New York Times*, March 10, 2018, available as https://www.nytimes.com/2018/03/10/world/asia/china-xi-jinping-term-limit-explainer.html.

national levels, thus making those paths arguably less secure than paths like the MIN path, where energy needs are supplied by resources that are often tapped by facilities that are more distributed and each smaller and less obtrusive (and polluting) than the other two paths.

Additionally, the nuclear facilities (including enrichment and reprocessing, nuclear power plants, and spent fuel transport) that are part of the BAU and, particularly, the MAX paths will require much more in the way of military security arrangements than the MIN path. These military security requirements increase military costs and enhance the possibility of conflicts between the military security apparatus and a population becoming accustom to greater social and economic freedoms. The impact of required nuclear sector security arrangements will be mirrored, to some extent, by the greater needs to secure supplies of oil and oil transport lanes; the needs to secure oil supplies will be highest in the BAU case.

4. Conclusions

4.1. What Alternative Scenarios of China's Nuclear Future Tell Us

As of 2017, China arguably sits at a point of decision, inflection, or possibly both in the evolution of its electricity generation system, its nuclear power future, and possibly its energy sector as a whole. Scenarios of nuclear power development in China, including those presented in this paper and many others, span the range from modest growth to 80-100 GWe by 2050 (from about 34 GWe today) to projecting growth to 300 to 400 or more GWe of nuclear power by 2050, with the beginnings of commercialization of fast reactor technologies. At the same time, transitions are occurring for China's coal-fired power fleet, with plans for future new capacity being rapidly scaled back, and smaller, less-efficient units being taken out of service. The rethinking of plans for expansion of coal-fired power is in part in response to progressively more stringent policies to address local and regional air pollution, but also in response to the two trends, particularly in recent years, of accelerating deployment of wind and solar power, and reduced growth in electricity demand, the latter particularly in comparison to the double-digit growth rates of recent decades.

Developing internally-consistent scenarios of China's electricity sector in general, and the nuclear energy sector within the electricity sector, with the different scenarios designed to serve the same needs for energy services in similar economic futures, provides a means to test policy directions. China could choose a path including rapid deployment of LWRs with spent fuel reprocessing, blending the resulting plutonium into MOx fuel and subsequent use of same in LWRs and, ultimately, in a fleet of fast reactors. Or an explicit or implicit policy (for example, through adjusting levels of power sector subsidies) could damp down the current nuclear build-out after those reactors currently under construction are built, such that additions in the decades after 2025 are modest, while aggressively encouraging energy efficiency and the development of solar and wind power, plus the supply-side changes (transmission systems, smart grids, and electricity storage, for example) that would be needed to maximize renewable energy usability. The estimates of future fuels use, costs, pollutant and waste emissions, and accompanying (typically) qualitative consideration of issues such as the relative political and social security ramifications that result from consideration of different future scenarios provide a way of testing and illustrating for policymakers the different ways of organizing the energy future of a nation.

4.2. Prospects for Meeting China's Future Energy Needs with Limited *or No* Increases in Nuclear Capacity and Proliferation-resistant Fuel Cycles

The comparison of the BAU, MAX, and MIN paths for the Chinese nuclear and electricity sectors, considered together with the existing body of China scenarios work described (in part) in section 3.2 of this paper, suggests that it will be possible for China to meet its economic development, GHG and air pollutant emissions reduction, and other goals without an extended and massive build-out of LWR capacity, and without expansions of uranium enrichment or reprocessing capacity beyond those projects now underway. Further, although the MIN case implies that China will not become a major exporter of nuclear power technologies, it also implies that China will continue along its current trend of being perhaps a dominant provider of renewable power systems. Nuclear sector costs in the MIN path are much lower, both on an aggregate basis and per unit of output, than in the other two paths, largely because of lower uranium, enrichment, and reprocessing costs. Although these costs are only a small part of the overall cost of providing energy services to the Chinese economy, indications from past experience and other studies is that the emphasis on energy efficiency and renewable energy will offer the opportunity for China to effectively address its environmental concerns without significant (if any) additional costs, relative to a reference path. Further, a path with less nuclear power and fewer front-end and back-end nuclear facilities will be arguably easier to deploy in a social and political sense, particularly as expectations for a stronger voice in how its future unfolds continue to grow among the Chinese citizenry.

The MIN scenario provides significant benefits over the other two cases in terms of plutonium production and stocks (transient and otherwise), and thus provides significantly lower risk of the proliferation of nuclear weapons. For the MIN path to become a reality, policy support for energy efficiency and renewable energy will need to take precedence over policy support for nuclear power. Trends in recent years, including the slow-down in reactor construction and re-thinking of nuclear safety regulations post-Fukushima, ongoing structural change in the Chinese economy away from heavy industry (and much-reduced growth in electricity needs), and exceedance of even the ambitions government targets for renewable energy all point toward the enhanced practicality of a low-nuclear path for the evolution of the Chinese energy sector. Acknowledgement of the benefits of the MIN path (or similar) for China by international political and trading partners, probably including international policies that encourage such a path and embracing energy paths of their own that de-emphasize nuclear power, enrichment, and reprocessing, would likely serve to encourage China to move toward a low-nuclear future.

China arguably is at a point in marketing its nuclear technologies abroad in which its technologies are not particularly competitive—as they are based on older U.S. and other Western designs—and it is facing a worldwide market for nuclear power that even a Russian reactor vendor has reportedly described as weak. If China's domestic market for nuclear power were to follow a trajectory more like the MIN path described above than the BAU or MAX paths, it seems likely that China's nuclear exports would be relatively de-emphasized. Building and maintaining the capabilities to export nuclear technologies, including to countries where nuclear weapons proliferation is a danger (or historical fact, as in Pakistan), will be technically and economically riskier and more difficult without a burgeoning domestic market to fall back on. As such, timely encouragement (including by example) of China by the U.S. and the rest of the international community to focus on non-nuclear technologies for power generation could contribute to influences already in play and induce China to focus its efforts on exporting technologies that do not carry a weapons proliferation threat.

How Bright is Taiwan's Non-Nuclear Future?

Jessica Drun and Alexa Tzu-An Lee

Overview

The nuclear energy debate in Taiwan has persisted for decades, enduring from when the island was under authoritarian-control to the present, under its vibrant democracy. Considerations in this debate run the gamut with other issues—from social justice and environmental protection to party politics and special interests. These issues are further exacerbated by the precarious geopolitics surrounding Taiwan—with its reliance on fuel imports, a blockade scenario in a Taiwan Strait contingency involving the Mainland could cripple not only its military, in terms of fuel for military platforms, but also the island itself and its ability to "hold its own" until support—if any—from the United States arrives.

Under such heightened sensitivities, policymakers in Taiwan—especially the current Democratic Progressive Party (DPP) administration under Tsai Ing-wen—have sought avenues to phase out nuclear power usage on the island. However, following through on campaign promises has proven easier said than done and the Taiwan government faces an uphill battle against a price-sensitive electorate, a long-term reliance on fossil fields, and a relatively immature, domestic renewable energy sector.

The stakes for Taipei are particularly high. If successful, Taiwan could act as a standard-bearer for countries in East Asia, particularly China but peripherally Japan and South Korea, that are seeking to expand their civilian nuclear power facilities—and with it, the potential to stockpile additional plutonium that could easily be converted into nuclear weapons or proliferate into the wrong hands. If Taipei is unable to reach its goals, then the message it sends to its counterparts in East Asia is that phasing out nuclear power is not politically viable and will likewise undercut efforts from activists in neighboring states advocating for such a change.

Given these considerations, this paper will explore the feasibility of ending nuclear power in Taiwan, with discussions on the island's current energy mix, the level of political support and will behind such initiatives, and potential policy moves that could create conditions more conducive to phasing out nuclear power.

Background

History and Political Backdrop

The most contentious—and most politicized—issue in Taiwan's energy debate has been nuclear energy. Taiwan's anti-nuclear movement was borne against the backdrop of increased global opposition to the energy source following the Chernobyl incident in 1986 and was further galvanized by a lack of government accountability and political channels to air grievances.[195]

Taiwan, at the time, remained under authoritarian rule and martial law, imposed by the Kuomintang (KMT). In the 1970s and early 80s, the KMT had commissioned three nuclear powers plants (See Maanshan, Chinshan, and Kuosheng in Figure 1), which were

195. Nicholas Freschi, "Taiwan's Nuclear Dilemma," *The Diplomat*, 15 March 2018, https://thediplomat.com/2018/03/taiwans-nuclear-dilemma/ (accessed 12 May 2018); Brian Hioe, "Anti-Nuclear Activism in Taiwan and Japan," *New Bloom Magazine*, 30 January 2015, https://newbloommag.net/2015/01/30/anti-nuclear-activism-in-taiwan-and-japan/ (accessed 12 May 2018)

managed by the state-owned Taiwan Power Company (hereafter, Taipower) but were imported as turnkey projects from Westinghouse and General Electric.[196] Taipower sought to build its own nuclear power plant in 1978, the Lungmen Nuclear Power Point or Nuclear Power Plant Four.[197] While the government did heed public concerns in regards to nuclear safety and froze plans on Nuclear Power Plant Four in the aftermath of Chernobyl, discontent persisted, as activists had intrinsically tied anti-nuclear issues with anti-authoritarian rule and felt that decisions regarding Taiwan's nuclear future were made at the whim of bureaucrats and without prior consultation with the citizenry—particularly local, and at times more vulnerable, populations near the sites of plants.[198]

---

196. Florence de Changy, "Taiwan presses ahead with home-built nuclear power plant despite safety fears," *The Guardian*, 14 February 2012, https://www.theguardian.com/world/2012/feb/14/taiwan-asia-pacific (accessed 12 May 2018); Sheng-Hisung Yu and Kwei-Luen Hwang, 2000, "Nuclear Power Development and Public Acceptance in Taiwan," Paper presented at: *12th International Workshop on Nuclear Public Information in Practice, Cankarjev Dom, Ljubljana, Slovenia, February 13-16, 2000, https://inis.iaea.org/collection/NCLCollectionStore/_Public/36/034/36034521.pdf*

197. "Key Dates in Taiwan's Anti-Nuclear Movement 1978-2019" ("台灣反核運動大事記 1978-2019"), Taiwan Environmental Information Center (環境資訊中心), 10 March 2018, https://e-info.org.tw/node/10598 (accessed 12 May 2018)

198. Li Hanlin (李翰林), "Anti-nuclear is anti-authoritarian, needs democracy, need participation!" (反核就是反獨裁 要民主 要參與！), Taiwan Environmental Information Center (環境資訊中心), 11 July 2016, https://e-info.org.tw/node/117002 (accessed 20 June 2018)

Figure 16: Graphic of Locations of Taiwan's Nuclear Power Plants "Locations of Nuclear Power Plants," Atomic Energy Council (Republic of China), 2005, https://www.aec.gov.tw/english/nuclear/article03.php

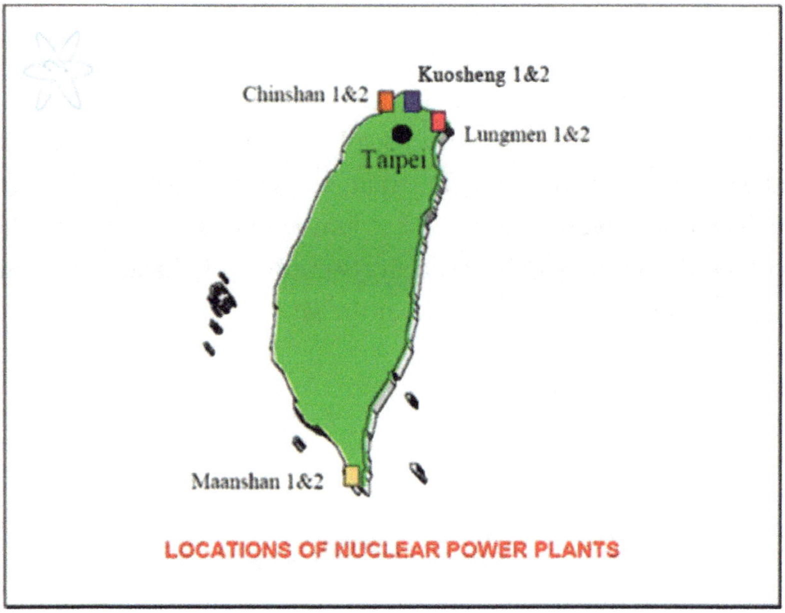

Accordingly, anti-nuclear activists found a natural ally—as well as a common cause in opposing the KMT–in Taiwan's democracy movement, to include members of what would later become the DPP.[199] When the DPP was formally established in 1986, the founding members incorporated environmental policies and a staunch anti-nuclear stance into their party platform.[200]

The first DPP President, Chen Shui-bian, was elected in 2000 during Taiwan's first transition of power—and with his inauguration came

199. Freschi, "Taiwan's Nuclear Dilemma"; Hioe, "Anti-Nuclear Activism in Taiwan and Japan"
200. Ming-sho Ho, "The Politics of Anti-Nuclear Protest in Taiwan: A Case of Party-Dependent Movement," *Modern Asian Studies* 37, no. 3 (2003): 683, http://www.jstor.org/stable/3876614

hopes from anti-nuclear activists that Taiwan would phase out nuclear power. During his campaign, Chen ran on a platform that called for the termination of the Fourth Nuclear Power Plant—which had begun construction in 1999.[201]

However, he was ultimately unable to deliver on these promises. Chen shelved the project when he entered office, but faced a KMT- and pan-blue held Legislative Yuan that opposed the move on the grounds of economic development and Taiwan's reputation in the eyes of foreign investors.[202] Further, the DPP arguably lacked the governance and institutional experience to navigate the bureaucracy and was outmaneuvered by the opposition.

In early February 2001, construction on Nuclear Power Plant Four resumed. Public discontent towards nuclear power ebbed and flowed for the next ten years, with high points coinciding with new developments or findings surrounding the Lungmen plant. For context, Nuclear Power Plant Four is particularly controversial, as compared to Taiwan's other three nuclear power plants. The announcement of its construction came around the time of the Chernobyl incident, when concerns over nuclear power were at an apex. The plant was also constructed piecemeal—over 800 separate projects were contracted out by Taipower and there appeared to be little coordination, which was assessed by one external party to "[bear] the risk of late and defective performance," at a cost significantly higher than those of the other three plants.[203]

---

201. Mark Landler, "Taiwan Ends Construction of Its 4th Nuclear Plant," *The New York Times*, 28 October 2000, https://www.nytimes.com/2000/10/28/world/taiwan-ends-construction-of-its-4th-nuclear-plant.html (accessed 20 June 2018); Robert Sutter, 2013 "Taiwan's Nuclear Debates and Implications," Paper presented at Carnegie Endowment for International Peace conference on Taiwan, China, and the United States, Washington, DC, October 3, 2013
202. Landler, "Taiwan Ends Construction of Its 4th Nuclear Plant"
203. Tsai Yaying, "2012 Shadow Report on Nuclear Safety," Wild at Heart Legal Defense Association, Taiwan, 2012,

Anti-nuclear sentiment in Taiwan hit an all-time high in the immediate aftermath of the March 11th earthquake and tsunami in Japan and the subsequent Fukushima Daiichi nuclear meltdown. Over two thousand protesters took to the streets.[204] This was followed by an April 30th "430 Sunflower No Nuke Action" rally, where over 16,000 individuals across the island marched in protest, calling for the government to stop construction on Nuclear Power Plant Four.[205] Every year after, protests have taken place on the anniversary of March 11th.[206]

In the face of these challenges and public debates, numerous narratives—both pro- and anti-nuclear energy—have been advanced and will be discussed subsequently. Further complicating the debate are the sensitive social issues that revolve around nuclear power and nuclear waste disposal. For example, given limited storage options, Taiwan has stored the bulk of its nuclear waste on Orchid Island near aboriginal communities who have consistently aired their grievances to the government and decried the impact these storage sites have had on their health, environment, and way of life.[207]

---

http://en.wildatheart.org.tw/sites/default/files/Shadow%20Report%20on%20Nuclear%20Safety_0.pdf?fbclid=IwAR3V4ZanArmPnnZUwdJ46Uk356ZV8FejfksCY6nGrcZC6RkL0W0FqvmlcwI (accessed 20 June 2018); "GcilA international dispute experience: Lungmen ABWR nuclear project in Taiwan," GcilA, 2018, http://www.gcila.org/experience/project-spotlights/lungmen-nuclear-power-plant/ (accessed 20 June 2018)

204. "Over 2,000 rally against nuclear plants in Taiwan," *Asia One News*, 20 March 2011, http://www.asiaone.com/News/Latest+News/Asia/Story/A1Story20110320-269104.html (accessed 20 June 2018)

205. Hermia Lin, "Protesters rally against nuclear energy in Taiwan," *Focus Taiwan*, 30 April 2011, http://focustaiwan.tw/news/asoc/201104300023.aspx (accessed 20 June 2018)

206. Freschi, "Taiwan's Nuclear Dilemma"

207. "Orchid Islanders protest ROC government nuclear waste policies," *Taiwan Today*, 1 December 2011, https://taiwantoday.tw/news.php?unit=10,23,45,10&post=18824 (accessed 20

Current Administration

The election of President Tsai Ing-wen in 2016 was viewed by many as the beginning of Taiwan's non-nuclear future. Tsai campaigned on a platform that included a "nuclear free homeland" by 2025.[208] Tsai and her campaign advocated for the suspension of Nuclear Power Plant Four and a no-extension of the license expirations of the other three nuclear power plants, as well as a move to increase power generation from renewables from 4% of total generation to 20% by 2025 and policy reform and coordination across major cities.[209]

Tsai secured the presidency alongside her party's first-time majority in Taiwan's Legislature, the Legislative Yuan (LY). A newly formed political party, the New Power Party (NPP, 時代力量)—which grew largely out of the Sunflower Movement of student activists—also won seats in the LY. The NPP also supports an end to nuclear power in Taiwan, and activists were optimistic that the combination of forces would bring about policies that set Taiwan down the path of phasing out nuclear energy.[210]

June 2018); Chang Tsun-wei and William Hetherington, "Lanyu residents demand removal of nuclear waste," *Taipei Times*, 2 August 2017, http://www.taipeitimes.com/News/taiwan/archives/2017/08/02/2003675764 (accessed 20 June 2018)

208. "The first meeting of the energy working group has ended, discussed the development of green energy and energy saving targets (新能源工作坊首場會議結束 提出開發綠能及節電具體目標)," Tsai Ing-Wen Campaign Website, 18 April 2015, http://iing.tw/posts/28?fbclid=IwAR3NbFVKyD1c2wVY6nWmTCUWRVjwt8lkDrxhm8vjAOuzLwZ631wJmwEfBsA (accessed 28 May 2019)

209. Ibid.

210. " 2016 Abolish Nuclear Energy March NPP will firmly take part (2016廢核大遊行　時代力量堅定參與)," New Power Party (時代力量), 12 March 2016, https://www.newpowerparty.tw/news/2016%E5%BB%A2%E6%A0%B8%E5%A4%A7%E9%81%8A%E8%A1%8C-%E6%99%82%E4%BB%A3%E5%8A%9B%E9%87%8F%E5%A0%85%E5%

However, the feasibility of such a goal is yet to be determined, with numerous factors at play—discussed subsequently—but to include economic and environmental considerations as well as issues with political will. To date, the government has already faced numerous roadblocks that have necessitated readjustments to its broader energy policy, including, most notably, a major blackout in the summer of 2017 that left 7 million households without electricity. [211]

Taiwan's Energy Overview

Despite the politics, the numbers surrounding Taiwan's energy usage paints a stark picture: the island still remains highly dependent on imported fossil fuels, as discussed in depth below.

Heavy Dependence on Energy Imports

Taiwan has few natural energy sources, with the island importing 98% of its energy supply in the form of fossil fuels and nuclear fuel, according to Taiwan government statistics for the year 2017.[212] The

---

AE%9A%E5%8F%83%E8%88%87 (accessed 28 May 2019); "DPP, NPP Show Up for Anti-Nuclear March, March Leader Urges the Two Parties to Take Action (民進黨、時力反核遊行現身 環團催兩黨快行動)," *Liberty Times (自由時報)*, 12 March 2016, https://news.ltn.com.tw/news/politics/breakingnews/1630373 (accessed 28 May 2019)

211. Jess Macy Yu, "Millions of People Were Left in the Dark as a Massive Power Cut Hit Taiwan," *Time*, 16 August 2017, http://time.com/4902500/taiwan-power-cut/ (accessed 28 May 2019)

212. Bureau of Energy, "Mobilizing Green Energy Transition and Industry Transformation through the Building of Green Energy Market," Republic of China Ministry of Economic Affairs, 12 February 2012, https://www.moea.gov.tw/MNS/english/news/News.aspx?kind=6&menu_id=176&news_id=76687 (accessed 20 May 2019); Sih Ting Jhou and Huei-Chu Liao, "Taiwan's Severe Energy Security Challenges," The Brookings Institution, 12

remaining 2% comes from a combination of hydro power, solar power, and geothermal power.[213] This is showcased in Figure 2 below, with the left side of the graphic showing the breakdown of Taiwan's energy supply and the right side indicating which sector uses what amount of energy and primarily from which sources. It is noteworthy that Taiwan's industries consume the largest amount of energy—tying the island's economic future to its ability to manage its energy policies.

Figure 9: "Energy Supply and Consumption Flowchart (2017)" by the ROC Bureau of Energy, Ministry of Economic Affairs[214]

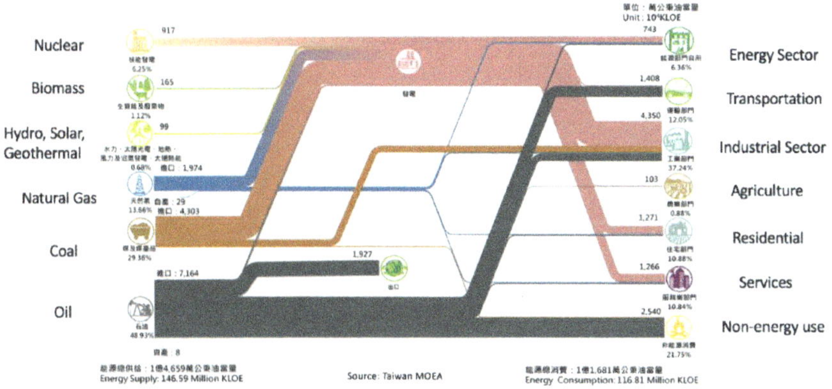

In more specific terms, the Bureau of Energy, under Taiwan's Ministry of Economic Affairs (MOEA), notes that this "total supply

September 2013, https://www.brookings.edu/opinions/taiwans-severe-energy-security-challenges/ (accessed 20 May 2019)
213. Sih Ting Jhou and Huei-Chu Liao, "Taiwan's Severe Energy Security Challenges"
214. "Energy Statistical Annual Report," Bureau of Energy, ROC Ministry of Economic Affairs, 1 August 2018, https://www.moeaboe.gov.tw/ECW/english/content/ContentLink.aspx?menu_id=1540 (accessed 27 May 2019)

of primary energy amounted to 146,635.3 10^3 kls of oil equivalent" in 2017, with "the total domestic final consumption of energy being 85,263.4 10^3 kls of oil equivalent."[215] Oil accounted for 48.93%, coal 29.36%, gas 13.66%, nuclear power 6.25%, biomass 1.12%, and a combination of hydro, solar, and geothermal amounted to 0.68%.[216]

Taiwan relies heavily on energy imported from the Middle East and other parts of the Indo-Pacific—with the bulk of these sources entering Taiwan through the Taiwan Strait, as summarized in Figure 3 below. This not only ties Taiwan's energy security—and by extension its economy—closely to international energy prices, but also highlights the risks posed to Taiwan in any cross-Strait contingency, as the People's Liberation Army Navy could viably cut off shipping vessels ferrying fossil fuels to the island.

215. Ibid.
216. Ibid.

Figure 10: Taiwan Energy Imports- Country Origins 2016. Translated by Author, Originally Graphics from the CTCI Foundation[217]

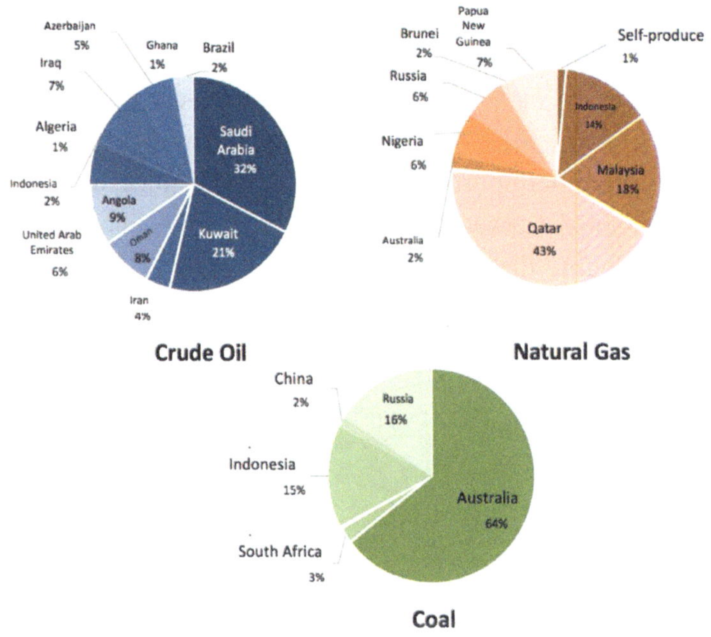

Further, high dependence on foreign energy sources also places a burden on Taiwan citizens, when considering energy economics and factoring in the content of the energy efficiency chart (Figure 4 below). The per capita electricity consumption in kWh shows that

---

217. Liang Qiyuan, Wang Wei, Liu Zhijun, Zheng Ruihe, Lu Yizen, and Guo Bozhen (梁啟源, 王釿鉻, 劉致峻, 鄭睿合, 呂易恂, 郭博堯), "2017 Taiwan Energy Performance vis-à-vis the World's and System Vulnerability (2017 台灣能源績效世界評比與系統脆弱度)," CTCI Foundation (財團法人中技社), December 2017, http://www.ctci.org.tw/media/4690/2017-07%E5%B0%88%E9%A1%8C%E5%A0%B1%E5%91%8A-%E5%8F%B0%E7%81%A3%E8%83%BD%E6%BA%90%E7%B8%BE%E6%95%88%E4%B8%96%E7%95%8C%E8%A9%95%E6%AF%94%E8%88%87%E7%B3%BB%E7%B5%B1%E8%84%86%E5%BC%B1%E5%BA%A6.pdf (accessed 20 May 2019)

Taiwan citizens have relatively stable electricity usage, with only minor increases each year, coming to about 10, 928 kWh in 2016—demonstrating steady growth. Yet, when this number is weighed against energy security indicators—which has seen per capita energy imports fluctuate significantly over the past twenty years—it becomes clear the key issue in regards to energy in Taiwan is not its electricity usage or demand, but rather how susceptible energy costs are to global market trends. The most illustrative example is the per capita energy import cost in 2012 due to historic highs in oil prices on the global market, with Taiwan's per capita energy import being nearly double the most recent rate in 2016.[218] Accordingly, while energy usage and demand remain stable, an undue burden is placed on Taiwan's people given instability in costs.

---

218. "Energy Statistics Handbook 2016 (能源統計手冊)", Bureau of Energy, MOEA (經濟部能源局), June 2017, https://www.moeaboe.gov.tw/ECW_WEBPAGE/FlipBook/EnergyStaHandBook/mobile/index.html#p=12 (accessed 20 May 2019), p. 10-12

Figure 11: Taiwan's Energy Efficiency Indicators and Energy Security Indicators, Chart by the ROC Bureau of Energy, MOEA[219]

2. 能源效率指標

Energy Efficiency Indicators

項目 Item / 年別 Year	年中人口數(千人) Mid-Year Population (1,000 Persons)	平均每人能源消費量(公升油當量/人) Per Capita Energy Consumption (LOE)	國內能源消費彈性值 Elasticity of Domestic Consumption	能源生產力(實質GDP/國內能源消費)(元/公升油當量) Energy Productivity (NT$/LOE)	能源密集度(國內能源消費/實質GDP)(公升油當量/千元) Energy Intensity (LOE/NT$1,000)	平均每人用電量(度/人) Per Capita Electricity Consumption (kWh)
1996	21,387.5	3,234.60	0.75	105.55	9.47	6,279.68
1997	21,577.0	3,352.06	0.74	107.13	9.33	6,640.90
1998	21,777.0	3,510.43	1.35	105.62	9.47	7,097.82
1999	21,952.5	3,652.32	0.73	107.47	9.30	7,331.40
2000	22,125.0	3,908.64	1.22	106.04	9.43	7,977.89
2001	22,278.0	4,099.71	-4.46	99.14	10.09	8,102.31
2002	22,396.5	4,258.96	0.80	100.22	9.98	8,495.36
2003	22,493.9	4,412.42	0.98	100.28	9.97	8,912.03
2004	22,575.0	4,587.07	0.67	102.37	9.77	9,297.40
2005	22,652.5	4,666.18	0.38	105.72	9.46	9,643.53
2006	22,740.0	4,739.39	0.35	109.52	9.13	9,936.97
2007	22,828.6	4,951.02	0.75	111.24	8.99	10,227.80
2008	22,904.8	4,794.59	-4.03	115.29	8.67	10,027.79
2009	22,979.4	4,685.81	1.25	115.74	8.64	9,604.61
2010	23,035.4	4,964.87	0.58	120.56	8.29	10,305.84
2011	23,082.9	4,899.97	-0.29	126.54	7.90	10,487.25
2012	23,151.2	4,875.39	-0.10	129.42	7.73	10,418.72
2013	23,216.0	4,991.97	1.22	128.82	7.76	10,557.13
2014	23,267.1	5,021.11	0.20	132.93	7.52	10,790.94
2015	23,320.1	4,996.08	-0.38	134.25	7.45	10,720.74
2016	23,369.4	4,998.36	0.17	135.89	7.36	10,928.00

Note: 1. Total primary energy supply = Indigenous + Imports - Exports - International marine bunkers - International civil aviation ± Change in stocks
2. Total final consumption = Industrial + Transportation + Agricultural + Services + Residential + Non-Energy Use
3. Total domestic consumption = Energy sector own use + Total final consumption

Source: 1. 「GDP」 is from Directorate-General of Budget, Accounting and Statistics, Executive Yuan
2. 「Mid-Year Population」 is from Dept. of Household Registration Affairs, MOI

3. 能源安全指標(續)

Energy Security Indicators (Cont.)

項目 Item / 年別 Year	能源進口值占總進口值比率(%) Value of Energy Imports/Value of Total Imports(%)	能源進口值占總出口值比率(%) Value of Energy Imports/Value of Total Exports(%)	能源進口值占名目GDP比率(%) Value of Energy Imports/Nominal GDP(%)	平均每人負擔能源進口值(臺幣元) Per Capita Energy Imports (NT$)	備用容量率(%) Percent Reserve Margin(%)	電力負載 Electricity Load: 尖峰負載(千瓩) Peak Load (MW)	電力負載 Electricity Load: 平均負載(千瓩) Average Load (MW)
1996	7.90	6.92	2.78	10,438	5.6	21,762	14,227
1997	7.47	6.91	2.84	11,466	11.0	22,237	15,097
1998	6.26	5.85	2.35	10,103	7.7	23,830	16,320
1999	7.26	6.52	2.65	11,850	12.5	24,206	16,639
2000	8.99	8.32	3.82	17,875	12.6	25,854	17,818
2001	10.81	9.35	3.94	17,951	13.2	26,290	18,043
2002	10.18	8.63	3.79	18,074	16.0	27,117	18,939
2003	11.44	9.84	4.67	22,761	14.6	28,594	19,841
2004	12.78	11.93	6.32	32,589	20.2	29,034	20,634
2005	15.81	14.66	7.77	41,462	16.3	30,943	21,651
2006	17.48	15.96	9.30	51,676	16.1	32,060	22,419
2007	19.71	17.67	10.77	63,240	16.2	32,791	23,043
2008	25.72	24.35	14.96	85,921	21.1	31,320	22,796
2009	21.65	18.69	9.80	55,295	28.1	31,011	22,101
2010	19.84	18.29	11.40	69,862	23.4	33,023	23,674
2011	22.63	20.83	13.37	82,885	20.6	33,787	24,320
2012	25.36	22.96	14.20	90,151	22.7	33,081	24,102
2013	23.57	21.03	12.74	83,608	17.5	33,957	24,364
2014	23.22	20.45	12.26	84,886	14.7	34,821	25,026
2015	15.81	13.14	7.08	50,871	11.5	35,248	25,012
2016	13.26	10.92	5.78	42,353	10.4	35,864	25,775

Taiwan's Electricity Supply and Demand

Statistics indicate that Taiwan's electricity energy consumption ranks it 14th in the world for energy per capita at 10,632 kWh/per person per year. With China ranking 1st, Japan ranking 5th, and South Korea 9th, Taiwan is the 4th highest in Asia on electricity consumption rankings.[220] According to data from Taiwan's Bureau of Energy, the total electricity consumption was 255,381 billion Ghw/year while generation was 261,114 billion Ghw/year.[221]

219. Ibid.
220. The World Factbook: Electricity Consumption 2016," Central Intelligence Agency, 2016, https://www.cia.gov/library/publications/resources/the-world-factbook/fields/253rank.html (accessed 20 May 2019)
221. "Energy Statistics Handbook 2016 (能源統計手冊)," p. 8

On the structural supply side, 2016 power generation constitutes 48.93% from oil (71,723.6000KLOE), 29.36% from coal plants (43,033,3000 KLOE), 13.66% from natural gas (20,030000KLOE), 6.25% from nuclear power (9,1691000KLOE), 1.12 % from biomass and waste (1,647000KLOE), and less than 1% from renewables and hydro power, as reflected in Figure 5:[222]

Figure 12: Structure of Energy Supply, Chart by Taiwan Bureau of Energy, MOEA[223]

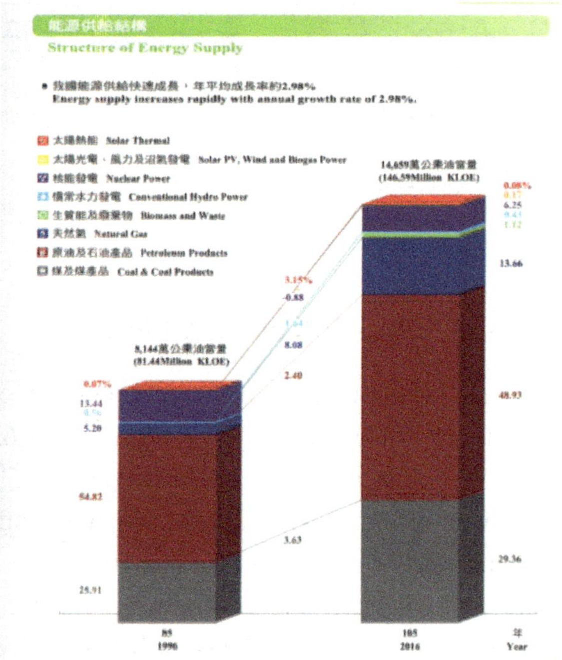

222. "Energy Statistics Handbook 2016 (能源統計手冊)," p. 24-25
223. "Energy Statistics Handbook 2016 (能源統計手冊)," p. 19

On generation costs, the Taiwan government has spent the most resources on gas generation, according to 2012 data from the National Policy Foundation of Taiwan. The purchase of coal for the Taiwan government cost $88 billion NTD/year, making up 30% of total costs, with oil representing 11%, and nuclear representing 2%. In comparison, the highest purchasing cost for natural gas was $168 billion NTD/year, constituting 57% of total costs.[224] Without counting factoring in facilities, equipment, and other sunk costs, the average cost of generating electricity in 2014 with oil was 5.58 NTD/kWh, with gas 3.57 NTD/kWh, with coal 1.05 NTD/kWh, and with nuclear 0.34 NT/kWh, as reflected in the chart below.[225]

Figure 13: Generation Costs by Fuel, Chart by the National Policy Foundation (translated by authors)[226]

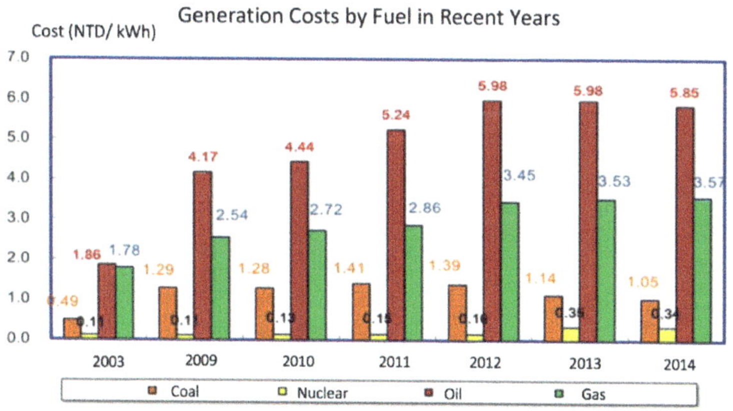

---

224. Tang Hui-Ling (唐慧琳), "Whose Fault for Rising Electricity Prices (電價上漲誰之過)," National Policy Foundation (國家政策研究基金會), 16 July 2013. http://www.npf.org.tw/printfriendly/12479
225. Ibid
226. Ibid

Industrial Sector Is a Driver for Economic Growth

Taiwan sits at the center of the global technology supply chain and electricity sustainability is crucial to the economic lifeline of the island. Figure 7 below shows the end-use by sector in Taiwan, with industry consuming 37.24% of total domestic consumption in 2016.[227] High-tech industries make up the largest sector of Taiwan's economy, with 13% of GDP growth driven by the semiconductor industry alone.[228] The Taiwan Semiconductor Manufacturing Co. (TSMC)—the largest contract chipmaker in the world—unilaterally drove up Taiwan's power usage by 33.6% over the course of five years, from 2012 to 2017.[229] This will only increase in the future given the requirements of the next generation of semiconductor production technology that is soon to be widely used.[230] As Taiwan's industries—and economic backbone—drive up demand for electricity, Taiwan's government will be in a tight position to ensure policies can support these growth.

---

227. "Energy Statistics Handbook 2016 (能源統計手冊)," p. 35
228. Chen Wei-Lau (陳威珞), "Industry Analysis: Semiconductor Trend 2018 (產業分析：半導體製造業發展趨勢2018年)," Taiwan Trend Research (台灣趨勢研究), 2016, https://www.twtrend.com/share_cont.php?id=59 (accessed 20 May 2019)
229. Chen Liang-Rong, "Can Taiwan Power TSMC's Dream?" *CommonWealth Magazine*, 25 August 2017. https://english.cw.com.tw/article/article.action?id=1663 (accessed 20 May 2019); Cheng Ting-fang and Lauly Li, "TSMC signals bottom of chip cycle as quarterly profits tumble," *Nikkei Asian Review*, 18 April 2019, https://asia.nikkei.com/Business/Companies/TSMC-signals-bottom-of-chip-cycle-as-quarterly-profits-tumble (accessed 20 May 2019)
230. Ibid.

Figure 14: Total Domestic Consumption (by Sector), Bureau of Energy, MOEA[231]

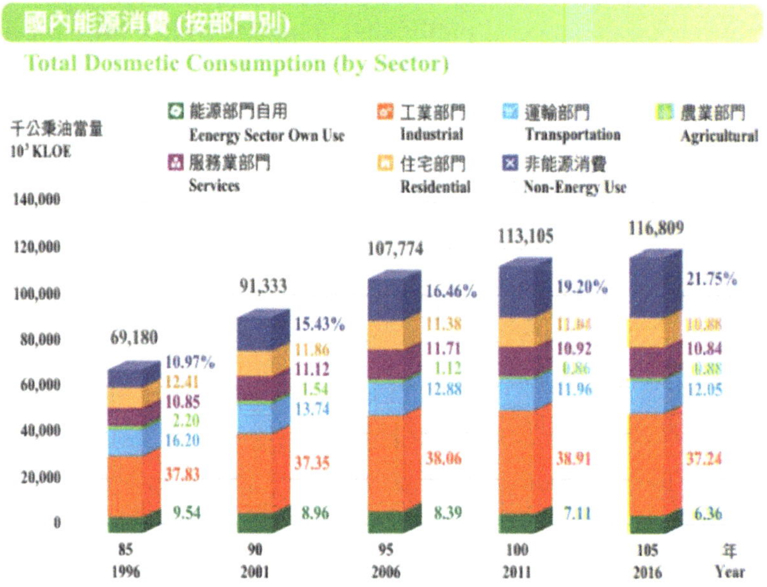

Monopoly Nature of Taipower

Taipower's monopoly over Taiwan's electricity infrastructure allows the island's consumers to have one of the lowest electricity costs in the world. According to a 2018 International Energy Agency (IEA) report, Taiwan's household electricity price of 2.47 NTD/kWh (roughly 7.8 cents USD) was the third lowest, behind Mexico and Malaysia, with the sixth lowest for industrial electricity rates, at 2.38 NTD/kWh (7.5 cents USD). [232] On the latter, Taiwan's electrical rates for industry is lower than Malaysia's of 2.7

231. "Energy Statistics Handbook 2016 (能源統計手冊)," p. 35
232. "Comparison of Each Country's Electricity Prices (各國電價比較)," Taipower (台灣電力公司), 3 October 2018, https://www.taipower.com.tw/tc/page.aspx?cchk=1b3221ee-37c3-4811-9d4d-a1bb215f33c8&cid=351&mid=213 (accessed 21 May 2019)

NTD/kWH, despite the fact that the Southeast Asian country has natural energy resources.[233] Accordingly, the artificially low prices of electricity in Taiwan highlights a critical issue in Taiwan's energy debate—that the general public and companies have little incentive to decrease their electricity usage and improve efficiency.

Though the electricity market was liberalized in 1995 and allowed for nine independent power producers (IPP), an "unfair agreement" between the nine IPPs and Taipower has existed that places financial burdens on Taipower. Purchasing prices are higher than Taipower can afford, with fixed-rates that do not reflect purchasing prices over time. In 2013, independent analysts calculated that the average electricity price was 2.98 NTD/kWh, but the price that Taipower paid the nine IPPs was higher, at an average 3.65 NTD/kWh—leading to imbalanced financials for Taipower in electricity generation. [234]

Thus, Taipower has struggled to fulfill its contracts with the IPPs and is losing money, though the nine IPPs have profited. According to the *Taipei Times*, Taipower lost almost $10 billion NTD (roughly $343 million USD) in the first half of 2017 due to cuts in electricity rates from April 2016—driving the company's accumulated losses to over $100 billion NTD (roughly $3.2 billion USD). [235] In 2017, the Taiwan Fair Trade Commission fined the nine IPPs $6 billion NTD (roughly $191 million USD) for price inflation, but the IPPs

---

233. "Comparison of Each Country's Electricity Prices (各國電價比較)"; "Malaysia- Oil and Gas," Export.gov, 19 July 2018, https://www.export.gov/article?id=Malaysia-Oil-and-Gas-Equipment (accessed 21 May 2019)

234. "The Secret to Why Taipower is Losing Money (台電虧損的秘密原因)," 21 August 2013. http://stock2012.pixnet.net/blog/post/155818181-台電虧損的秘密原因 (accessed 21 May 2019)

235. "Taipower's Accumulated Losses Top NT$100 Billion," *Taipei Times*, 18 July 2017, http://www.taipeitimes.com/News/biz/archives/2017/07/18/2003674762 (accessed 20 May 2019)

won the follow-on case with Taipei's High Administrative Court, keeping Taipower at a disadvantage. [236] Accordingly, Taipower is ill-positioned to upgrade any infrastructure. The company is already tasked with electricity generation, transmission, and distribution across the board, leading to low incentives to improve efficiency by increasing its capacity and addressing challenges posed by the energy market.

In 2016, Taipower's peak load grew by 1.75% as compared to 2015, coming to 35,864 MW—the highest in its history.[237] Taiwan's operating reserve—or its short-term generation capacity in the case of a disruption—fell below 6%, or the rate that triggers a power supply warning, 80 days in 2016, 104 days in 2017, and 29 days in 2018. [238] In August 2017, Taiwan's Minister of Economic Affairs resigned over a major blackout that Taipower attributed to "human error." [239] During the power outage, Taiwan's operating reserve fell to less than 2%.[240] The five-hour ordeal cost an estimated loss of $87

---

236. Wang Ji-Yiu (王己由), "9 IPPs Win, will not be Fined 6 Billion NTD (9民營電廠勝訴 60億罰單免繳*)," Commercial Times* (工商時報), 26 May 2017, https://m.ctee.com.tw/focus/cjzc/150485 (accessed 20 May 2019)
237. Lin Yi-Chun (林怡均), "Interpreting Taiwan's 2016 Energy Situation (解讀臺灣2016能源情勢)," NTU Risk Management and Policy Research Center (台大風險社會政策研究中心), 29 March 2017, http://rsprc.ntu.edu.tw/zh-tw/m01-3/energy-transformation/627-0330_tw-energy-status-2016 (accessed 20 May 2019)
238. "History Electricity Supply Demand Information (過去電力供需資訊)," Taipower (台電), 31 May 2019. https://www.taipower.com.tw/tc/page.aspx?mid=210&cid=341&cchk=ae51e573-e600-4a77-bc48-9e8d8f1b70c1 (accessed 1 June 2019)
239. Lawrence Chung, "Taiwanese Minister Resigns After Error at Power Plant Plunges 668000 Homes into Darkness,"15 August 2017, https://www.scmp.com/news/china/society/article/2106935/taiwanese-minister-resigns-after-error-power-plant-plunges-668000 (accessed 20 May 2019)
240. Tim Ferry, "Operating on the Edge: Huge Pressure on Taiwan's Power Supply,"24 October 2017, https://international.thenewslens.com/article/81604 (accessed 1 June 2019)

million NTD (approximately $3 million USD), showcasing just how critical energy security is to Taiwan's economy.[241]

Taiwan Energy Future and Recommendations

Tsai's Plan for 2025

The current administration is aware of these risks. In 2016, President Tsai released a plan with the target coal of phasing out nuclear by 2025 by decreasing coal's share in power generation from 45% to 30% and raising the share of natural gas from 32% to 50%.[242] To attract foreign and domestic investment, the government has released numerous policy initiatives, including ones on green finance and reforming Taipower. Further, the Taiwan government is laying the groundwork for upgrading the island's existing electricity infrastructure and expanding the grid to foster greater connectivity across the whole island, including too distant, rural communities that are not part of the current grid system.[243] At the same time, the

241. "Taiwan Power Outage Affected 151 Companies, caused $3 Million In Damages," *Reuters*, 17 August 2017, https://www.reuters.com/article/us-taiwan-power-outages/taiwan-power-outage-affected-151-companies-caused-3-million-in-damages-idUSKCN1AX0S3 (accessed 1 June 2019)

242. Robin Hicks, "Is Taiwan's 20 percent renewable energy target a political pipe dream?" Eco-business, 19 October 2017, https://www.eco-business.com/news/is-taiwans-20-per-cent-renewable-energy-target-a-political-pipe-dream/ (accessed 1 June 2019); Keoni Evertington, "Taiwan sets goal of 20% electricity from renewables by 2025," *Taiwan News*, 10 March 2017, https://www.taiwannews.com.tw/en/news/3113567 (accessed 1 June 2019)

243. "Tsai Ing-wen: Outlook of Taiwan's Renewable Energy Is Optimistic; Nuke-free Homeland's Energy Supply Will Be Sufficient," EnergyTrend, 1 October 2018, https://m.energytrend.com/news/20181001-12497.html (accessed 1 June 2019); Christine Chou, "FEATURE: Taiwan takes first step in power sector reform," *Asia News* Network, 18 December 2016, http://www.asianews.eu/content/feature-taiwan-takes-first-step-power-sector-reform-35468 (accessed 1 June 2019); Judy Lin, "Taiwan gov't draws investors with new green financing plans," *Taiwan News*, 10 April 2017, https://www.taiwannews.com.tw/en/news/3137332 (accessed 1 June 2019);

Tsai administration also tied these energy goals with its efforts to reduce the risks of climate change and lessen carbon emissions, with some analysts seeing the move as a way to attract greater international investment from companies looking to minimize their carbon footprints. Others, however, view the move as overly ambitious and potentially detrimental to Taiwan's power system, given low reserve margins.[244]

The Tsai administration has initiated numerous plans and policies towards these goals. The below chart showcases examples of forward-progress in the Tsai administration's efforts to reach its energy targets:

Year	2016	2025
	Nuclear 12%	Nuclear 0%
Developments	- In December 2018, the first reactor of Nuclear Power Plant 1 entered its decommission phase after 40 years of operations.[245] - In February 2019, Taipower dismissed calls to activate Nuclear Power Plant 4 amidst electricity shortages and the Ministry of Economic Affairs asserted that "there will be no extension or restarting of nuclear power plants in Taiwan."[246]	
	Natural Gas 32%	Natural Gas 50%

"Launch of 'Energy for the Next Generation' report," European Chamber of Commerce Taiwan, 30 May 2018, https://www.ecct.com.tw/launch-of-energy-for-the-next-generation-report/ (accessed 1 June 2019)
244. Edward White, "Taiwan green shift defies energy security fears," *Financial Times*, 2 January 2018, https://www.ft.com/content/6b429f2e-e3c6-11e7-97e2-916d4fbac0da (accessed 1 June 2019)
245. "First Taiwanese reactor to enter decommissioning," World Nuclear News, 5 December 2018, https://www.world-nuclear-news.org/Articles/First-Taiwanese-reactor-to-enter-decommissioning (accessed 1 June 2019)
246. "Taipower rules out activation of Fourth Nuclear Power Plant in face of shortages," *Taipei Times*, 2 February 2019, http://www.taipeitimes.com/News/biz/archives/2019/02/02/2003709080 (accessed 1 June 2019)

Developments	- Taiwan's CPC Corporation and Taipower are in the process of acquiring permits to develop Taiwan's 3rd LNG terminal.[247] - Taipower is in the early phases of commissioning Taiwan's 4th LNG terminal in Hsieh-Ho and has conducted a feasibility study for a power plant at that site.[248]	
	Coal 46%	Coal 30%
Developments	- Taipower has announced that it is not planning to develop any new coal-fired plants or generators.[249] - In October 2018, the government terminated plans to reopen and expand the controversial Shenao coal-fired power plant.[250] - In April 2019, Taipower was fined $20 million NTD for a coal-fired power discharging wastewater pollution—the highest charge ever levied against a state-owned enterprise in Taiwan.[251]	
	Solar PV 1.7 GW	Solar PV 20 GW
Developments	- In 2019, Executive Yuan released the "Green Energy Rooftop Participation Plan" to accelerate	

247. Florence Tan, "Taiwan's CPC to start building third LNG terminal by mid-2019," *Reuters*, 12 March 2019, https://www.reuters.com/article/us-ceraweek-energy-cpc-taiwan/taiwans-cpc-to-start-building-third-lng-terminal-by-mid-2019-idUSKBN1QT2MT (accessed 1 June 2019)
248. Mike Corkhill, "Taiwan Power maps out Hsieh-Ho LNG import terminal plan," LNG World Shipping, 16 July 2018, https://www.lngworldshipping.com/news/view,taiwan-power-maps-out-hsiehho-lng-import-terminal-plan_53532.htm (accessed 1 June 2019); Jessica Jaganathan, "Taipower expects to start LNG imports from 2023," *Reuters*, 7 March 2019, https://www.reuters.com/article/us-singapore-lng-conference-taipower/taipower-expects-to-start-lng-imports-from-2023-idUSKCN1QO137 (accessed 1 June 2019)
249. "Taiwan Power Co. respects referendum results but will follow government policy," *Taiwan News*, 25 November 2018, https://www.taiwannews.com.tw/en/news/3583184 (accessed 1 June 2019)
250. Teng Pei-ju, "Government to scrap Shenao power plant project: Taiwan premier," *Taiwan News*, 12 October 2018, https://www.taiwannews.com.tw/en/news/3550883 (accessed 1 June 2019)
251. Natashi Li, "Taichung Power Plant fined NT$20m," *Taipei Times*, 16 April 2019, http://www.taipeitimes.com/News/biz/archives/2019/04/16/2003713443 (accessed 1 June 2019)

	- meeting the rooftop PV goal of 3 GW and the land PV goal of 17 GW.[252] - Taipower announced in May 2019 that it plans to develop the island's largest solar power facility in Southern Taiwan.[253]	
	Wind 692 MW	Wind 6.7 GW (1.2 GW land-based, 5.5 GW offshore)
Developments	- Development for Phase 2 of Formosa 1—Taiwan's first offshore wind project—started in early 2019.[254] - In September 2018, Taiwan released its "Wind Power Four Year Plan" to expedite foreign investment and ensure regulatory cohesion.[255] - In January 2019, Taiwan's MOEA decreased the feed-in tariff (FiT) rate for offshore wind projects by 6%.[256] - In April 2019, Ørsted finalized investment for 900 MW of offshore wind projects in Changhua. [257]	

252. "Green Energy Rooftop Participant Plan(綠能屋頂全民參與計畫), Executive Yuan (行政院), 2 January 2019, https://www.ey.gov.tw/Page/5A8A0CB5B41DA11E/e9cb7d49-3982-4f9d-8cfe-8f7096df3acc (accessed 1 June 2019)
253. Chang Jung-hsiang and Frances Huang, "Taipower to build nation's largest solar power facility in Tainan," *Focus Taiwan*, 22 May 2019, http://focustaiwan.tw/news/aeco/201905220016.aspx (accessed 1 June 2019)
254. "Foundations arrive for Formosa Phase 2," 4C Offshore, 10 May 2019, https://www.4coffshore.com/news/foundations-arrive-for-formosa-phase-2-nid13626.html (accessed 1 June 2019)
255. Wind Power Four Year Plan (風力發電四年計畫), Executive Yuan (行政院), 10 September 2019, https://www.ey.gov.tw/Page/5A8A0CB5B41DA11E/ef93b5c1-85ea-4b5f-ac55-f460d9204258 (accessed 1 June 2019)
256. "Taiwan Cutes 2019 Offshore Wind FiT by 6%," Renewables Now, 30 January 2019, https://renewablesnow.com/news/taiwan-cuts-2019-offshore-wind-fit-by-6-641188/ (accessed 1 June 2019)
257. "Ørsted Greenlights 900 MW Taiwanese Offshore Wind Farm," OffshoreWindBiz, 30 April 2019, https://www.offshorewind.biz/2019/04/30/orsted-greenlights-900mw-taiwanese-offshore-wind-farm/ (accessed 1 June 2019)

However, there have also been numerous setbacks since the Tsai administration took office, including restarting reactors at nuclear power plants to accommodate for energy demand surges and moving forward on projects despite concerns about the environment.[258]

Energy as a Political Issue

Yet, the crux of Taiwan's energy issues will always come down to politics. As discussed above, nuclear energy was borne of Taiwan's democracy movement and has become subject to partisan whims and debates. For the Tsai administration, this all came to the fore in November 2018 during the referendum that coincided with Taiwan's nine-in-one local elections. Three of the questions on the referendum addressed energy issues and were as follows:

- "Case 7: Do you agree 'To reduce by 1% year by year' the production of Coal-Firing Thermal Power Plants?
- Case 8: Do you agree to the establishment of an energy policy to 'Stop construction and expansion of any Coal-Firing Thermal Power Plants or generator units (including the Shen Ao Power Plant currently under construction)'?
- Case 16: Do you agree to repeal Article 95 paragraph 1 of The Electricity Act: 'The nuclear-energy-based power-generating facilities shall wholly stop running by 2025'?"[259]

---

258. Brian Hioe, "Tsai Administration Quietly Approves of Nuclear Reactor Restarts: Outrage to Follow?" *New Bloom*, 15 June 2017, https://newbloommag.net/2017/06/15/nuclear-restarts-tsai/ (accessed 1 June 2019); "Environmental assessment passed and the 3rd LNG terminal is allowed to begin construction (環評案通過中油第三天然氣接收站可建置)," CNA (中央社), 8 October 2018, https://www.cna.com.tw/news/firstnews/201810085005.aspx (accessed 1 June 2019)

259. "Case Referendums," Central Election Commission (ROC), November 2018,

All three were adopted. Many view the results as reflective of Taiwan citizens' concern for clean air (Cases 7 and 8), with the belief that a rapid reduction in nuclear energy (Case 16) would come with burning more fossil fuels and more carbon emissions.[260] In effect, the outcome was not necessarily in opposition to phasing out nuclear energy, just at the proposed pace. Interviews with voters in favor of Case 16 found that there were likewise worries over the reliability of renewable energy sources, the high costs of these sources, and potential economic ramifications.[261] Others, however, have attributed the outcome to a disparity in grassroots efforts, to disinformation campaigns, and to insufficient debate on the matter given the media focus on other referendum topics such as same sex marriage.[262] Opponents purport that the referendum on nuclear energy does not accurately represent the views of the Taiwan people.[263]

In the aftermath, the Legislative Yuan—in accordance with the referendum results—removed Article 95, or the provision that nuclear power plants terminate operation by 2025—from Taiwan's

---

https://web.archive.org/web/20181125031636/http://referendum.2018.nat.gov.tw/pc/en/00/00000000000000000.html (accessed 1 June 2019)

260. Chris Horton, "Taiwan Asked Voters 10 Questions. It Got Some Unexpected Answers." *The New York Times*, 26 November 2018, https://www.nytimes.com/2018/11/26/world/asia/taiwan-election.html (accessed 1 June 2019)

261. Michael Shellenberger, "Pro-Nuclear Activists Win Landslide Electoral Victory In Taiwan," 24 November 2018, https://www.forbes.com/sites/michaelshellenberger/2018/11/24/pro-nuclear-activists-win-landslide-electoral-victory-in-taiwan/#24daf9d629a0 (accessed 1 June 2019)

262. Nick Aspinwall, "How Direct Democracy Went Nuclear in Taiwan," *The Diplomat*, 18 January 2019, https://thediplomat.com/2019/01/how-direct-democracy-went-nuclear-in-taiwan/ (accessed 1 June 2019)

263. Ibid.

Electricity Act.[264] Notwithstanding, the Tsai administration has reaffirmed its commitment to its stated target of 2025.[265]

In essence, despite overall shared concerns over the environment and common end goals of decreasing carbon emissions, the Taiwan electorate disagreed with the pace and target goal of the Tsai administration's "nuclear free homeland by 2025." Yet, the outcome may have implications beyond the immediate referendum and the likely extension of the nuclear phase-out timeline, as it served as a major blow to the Tsai administration and one of her major policy goals. This can be perceived by both political opponents and the general Taiwan population as representative of Tsai and the DPP's ability to deliver on campaign policies and to govern, with consequences for election results—and future energy policies—down the line.

Policy Recommendations

Given Taiwan's energy realities and political dynamics, as discussed in detail above, the prospect of phasing out nuclear energy appears to be grim. The following policy options provide an overview of possible paths forward:

---

264. Sean Lin, "Provision on halting nuclear power plants removed," *Taipei Times*, 8 May 2019, http://www.taipeitimes.com/News/taiwan/archives/2019/05/08/2003714760 (accessed 1 June 2019)

265. "Taiwan government maintains nuclear phase-out," World Nuclear News, 1 February 2019, https://www.world-nuclear-news.org/Articles/Taiwan-government-maintains-nuclear-phase-out?fbclid=IwAR3ISEMXvA3-d0f77YKJg4PgqBQZ5zPNB8AzS45eZrcV-GoSBWALAG-lB1w (accessed 1 June 2019)

Demand Management Innovation

According to its spokesperson, Taipower admitted that the company is behind on its electricity generation, due to rising temperatures and increased electricity consumption over the past two years, which resulted in the low margin of reserve power.[266] In 2017, the Taiwan government restricted all government offices from using air-conditioning between 1 p.m. and 3 p.m. This resulted in some criticism, with opponents viewing the approach as a band-aid that covered the more fundamental issues at play, all while refusing to address the reality of the situation.[267] Given the limited options Taiwan has on the table on this front, demand-side management is a crucial component of a longer-term policy that the island must look into. Consumers should also be incentivized to act and contribute to any proposed energy saving schemes as early as possible.

Examples and best practices from other countries that have focused on demand-side management could serve as models for Taiwan to emulate. For example, the United States' Automated Demand Response system has proven to be a successful tool for demand management, allowing consumers to fully rely on the Internet to auto-manage their electricity usage and without time-consuming human interactions.[268] demand response programs are being used by system planners and operators across the world as a means of balancing supply and demand, while time-based rates offer

---

266. Peng Xingzhu (彭杏珠), "Largest Electricity Shortage will hit in 2025, with Costs of Electricity up 45% (史上最大缺電潮來襲 2025年電力成本恐漲 45％)," Anue (鉅亨網) 7 May 2018, https://news.cnyes.com/news/id/4114296 (accessed 3 May 2019)
267. Liu Lizen(劉力仁), "Executive Yuan Plans to Walk through Difficulties Together (政院盼共體時艱)", *Liberty Times Net* (自由時報), 3 August 2017, https://news.ltn.com.tw/news/life/paper/1123915 (accessed 20 May 2019)
268. Lawrence Berkeley National Laboratory, "Costs to Automate Demand Response-Taxonomy and Results from Field Studies and Programs," November 2015, https://gig.lbl.gov/sites/all/files/drrc_final_report_taxonomy.lbnl-1003924.pdf (accessed 20 May 2019)

consumers greater awareness of electricity pricing, peak rates, and other relevant information to manage energy usage from the demand side. According to a Federal Energy Regulatory Commission report:

> [T]here are two types of demand response programs – one that induces the consumer to alter behavior by offering certain incentives (incentive-based programs), and the other that offers price signals to make customer change her usage pattern (time-based programs) and the former has proved to be more popular in the market. [269]

Thus, the key to success depends on the diversity of utility programs and the amount of consumer participation. For example, Detroit Edison in Michigan reported a 3,000 MW demand reduction from its time-of-use program for commercial and industrial consumers due to attractive rates.[270] Similarly, in Maryland, Baltimore Gas and Electric reported increased participation in their scheme by incorporating greater amounts of residential participation.[271]

Australia also introduced various demand management programs in 2018.[272] In different parts of the country, there are various systems already in place to reward customers for using distributed energy resources to provide services to the grid, as well as voluntary

---

269. Federal Energy Regulatory Commission, "2018 Assessment of Demand Response and Advanced Metering," November 2018, https://www.ferc.gov/legal/staff-reports/2018/DR-AM-Report2018.pdf (accessed 3 May 2019)
270. Michigan Public Service Commission, "Overview of Demand Response Programs in Michigan", 21 September 2015, https://www.michigan.gov/documents/energy/MPSC_Demand_Response_Presentation_September21_References_502211_7.pdf (accessed 20 May 2019)
271. Coley Girouard, "Behavioral Demand Response Gives Baltimore Gas and Electric a Business Reason to Reduce Peak Usage," Utility Dive, 28 January 2019, https://www.utilitydive.com/news/behavioral-demand-response-gives-baltimore-gas-and-electric-a-business-reas/546895/ (accessed 20 May 2019)
272. Irina Umback, "Demand management is the new black," Energy Networks Australia, 22 March 2018, https://www.energynetworks.com.au/news/energy-insider/demand-management-new-black (accessed 5 May 2019)

summer saver programs that rewards customers for reducing electricity consumption when asked by the network.[273]

The abovementioned experiences and examples could be beneficial case studies for Taiwan at a time when the island needs greater integration of solar into its grid system—which can potentially be combined with future, long-term infrastructure planning. Demand management presents a solution to build traditional network infrastructure to meet normal and peak demand while also addressing the fundamental problem surrounding a lack of electricity.[274] It can also address risks associated with equipment failure, delay, and replacement of aging assets by supplying information from the demand side.

However, there are several key challenges that Taiwan currently faces in fully developing a demand response system. First, there are few available options for both metering infrastructure and grid modernization. Accordingly, Taipower has borne the brunt of the responsibility in providing and developing infrastructure, as well as operating utility programs for customers—a stark contrast from the United States and Australia examples, where competition and diversity are allowed and encouraged. Taipower thus is entirely responsible in distributing automated metering infrastructure (AMI) to 3 million households within the next five years, an ambitious goal that the Bureau of Energy set in 2016.[275] However, the number of households that are aware of—let alone have participated in—smart metering programs in Taiwan are very low at the moment, due to a

273. Dan Silkstone, ARENAWIRE, "The Australian Summer is Coming and so is Demand Response", 8 September 2017, https://arena.gov.au/blog/demand-response-2/ (accessed 20 May 2019)
274. Irina Umback, "Demand Management is the New Black"
275. Chen Yuzen (陳郁仁), Taipower AMI (台電AMI), "The Government Passes Plans for 3 Millions Households to Change to AMI in 8 Years (政院通過300萬戶8年後換智慧電表)", 23 September 2016, http://www.ami-tech.tw/政院通過-300萬戶8年後換智慧電表/ (accessed 20 May 2019)

lack of incentives, but also to insufficient promotion of such programs. There are currently fewer than 10 programs with a combination of rates and providers that customers can choose from. Accordingly, Taipower needs to better promote and tailor program options for its diverse set of customers, which vary from residential to industrial users. During trial or early phases, Taipower could consider implementing a voluntary program in the summer with strong incentives and allowances to attract greater participation then adjust rates overtime.

Another approach worth considering is forming public-private partnerships with regional partners to allow localized methods to meet different electricity demands and coordinate management across Taiwan.

For example, Japan's Ministry of Environment has partnered with Opower, Oracle's utility group, to develop an energy savings program that resulted in an average of two percent less energy used in households that participated in the program compared to those that did not.[276]

These energy savings solutions are designed to inform and motivate consumers to better control their home energy consumption and cost while allowing utilities to meet regulatory requirements, reduce cost, and improve customer satisfaction. Given the limited strengths of Taipower, attracting private investment and expertise are crucial in energy management, and Taiwan should consider more partnerships in this space to unleash the potential in demand-side management.

---

276. "Consumers Reach Nearly 23 TWh of Energy Savings with Oracle Utilities Opower," Oracle, 19 February 2019, https://www.oracle.com/corporate/pressrelease/energy-savings-opower-021919.html (accessed 5 May 2019)

Accelerate Taiwan's Natural Gas Capacity

Based on Taiwan's past experiences, it is and would be difficult to cut coal and nuclear power usage at the same time. Natural gas could serve as a transitional fuel to pave the way for the renewable energy and renewable energy infrastructure to mature, especially if enough capacity can be built during this phase. In recent years, the primary reason for the increased use of natural gas in electricity production around the world is its falling price point. The relatively affordable price of natural gas will make it an economically attractive alternative fuel source as regulations on CO2 and other types of emissions are tightened. However, there are many roadblocks to this option, such as a lack of natural gas infrastructure and environmental impact considerations.

With the current 1.5 GW capacity for gas facility in 2016, Taiwan would need to build up to 2.7 GW capacity by 2025.[277] Given that the cost of gas may increase over the years, the Taiwan government will need to determine avenues to increase revenues and secure adequate gas sources. On liquefied natural gas (LNG), with the current 14 million metric tons in 2016, the government will need to build its capacity up to 29 million metric tons (with five plants/ with the addition of three plants) by 2025, though Taiwan has been falling behind schedule.[278]

The primary reasons for the delays are environmental assessments. The proposed gas sites face local opposition to the construction of gas facilities and relevant infrastructure in those areas. As such, the government will need to ensure that the expansion of natural gas infrastructure is built with sufficient oversight and sound environmental safeguards and undergo thorough assessments.

---

277. "Energy Transition Roadmap (能源轉型路徑規劃)," Republic of China Ministry of Economic Affairs (中國民國經濟部), 13 June 2017, https://www.ey.gov.tw/File/FE659A8C337D69EB?A=C (accessed 5 May 2019)
278. Ibid.

Recently, in October 2018, the environmental assessment for the controversial Third LNG Terminal in Taoyuan passed.[279] This project is controversial because the natural gas facility sites are near preserved lands and algae reefs, which could have potential effects on the local ecosystem, complicating the environmental assessment procedures. Even though revised plans agreed to reduce the affected area to a tenth of the originally proposed size, local opposition remains. Despite the fact that that the Deputy Minister stepped down due to his opposition against the project, the final vote on environmental assessment procedures passed with seven votes amidst an imbalance in government and civilian representation.[280]

Accordingly, in order to accelerate Taiwan's natural gas capacity, the government needs to establish a clear and balanced framework, assessing energy security and environmental impact and providing alternatives and remedies for local residents, in consultation with these individuals and environmental NGOs to find common ground. Otherwise, future gas projects—such as the two under way in Keelung and Taichung—may face similar opposition and backlash.[281] The importance of such deliberations is compounded by the fact that expanding gas infrastructure requires securing land rights, which involves diverse stakeholders with oftentimes conflicting interests, while also touching on sensitive social and environmental justice issues.

---

279. "LNG Terminal Construction Plan Passes Environmental Impact Evaluation," *Focus Taiwan*, 8 October 2018, http://focustaiwan.tw/news/aeco/201810080019.aspx (accessed 20 May 2019)
280. "The Darkest Environmental Impact Assessment in History! A minute look into the quarrel of the Kwun Tong Environmental Impact Assessment! (環評史上最黑暗！一分鐘看觀塘環評案到底在吵什麼？)," *UDN (聯合新聞網)*, 9 October 2018, https://udn.com/news/story/12525/3411584, (accessed 20 May 2019)
281. "Taiwan's CPC to start building third LNG terminal by mid-2019," *Reuters*, 12 March 2019 https://www.reuters.com/article/us-ceraweek-energy-cpc-taiwan/taiwans-cpc-to-start-building-third-lng-terminal-by-mid-2019-idUSKBN1QT2MT (accessed 20 May 2019)

Notwithstanding, there are examples that the Taiwan government may follow that effectively balance LNG development with sound environmental considerations. For example, Australia Pacific has conducted thorough environmental assessments in Queensland ahead of a natural gas project. [282] In 2010, the assessment found that the natural gas scheme upheld high standards, after undergoing studies of the local water terrain and with extensive stakeholder engagement that outlined a social impact management plan and reporting guidelines based off of local government requirements.[283] The environmental assessment study was approved within two years, with follow-up, annual evaluations, and the project continued successfully.[284] The Natural Resources Defense Council (NRDC) has also laid out crucial groundwork on the environmental assessment process, to include transparency, and a period for comments and screening, among others, that Taiwan could replicate in, or add to, its environmental assessment laws.[285] In short, for Taiwan to ensure that its natural gas capacity can expand vis-à-vis environmental considerations, structural changes are needed within the island's environmental assessment laws to allow sustainable growth in both gas capacity and the environment.

---

282. "Australia Pacific LNG Project: Coordinator-General's report on the environmental impact statement," Queensland Government, November 2010, https://www.aplng.com.au/content/dam/aplng/reports/2010/CG-Report-Australia-Pacific-LNG-Project.pdf (accessed 5 May 2019)
283. Ibid.
284. Ibid.
285. Natural Resource Defense Council, "Key Components and Best Practices for Environmental Impact Assessments," United Nations, 24 August 2016, https://www.un.org/depts/los/biodiversity/prepcom_files/NRDC_HSA_EIA_Brief_PrepCom2.pdf (accessed 10 May 2019)

Increase Cleaner Coal Generation in the Short-term

As mentioned above, coal is the biggest source of Taiwan's electricity generation. The 2016 capacity of goal generation was 11 GW, with the stated goal of reaching 14 GW by 2025. In terms of energy mix, this plan will proportionally decrease from 45% to 30% in the long term.[286] Taiwan's thermal power plant is located in the Taichung area, and coal and gas constitute 45% of its generation sources, respectively. Taichung's thermal plant is the world's single largest thermal plant and its total electricity generation has encompassed 2% of Taiwan's total annual power production.[287] However, high production brings with it environmental concerns and the Taichung municipal government should work in collaboration with Taipower to determine means of raising overall environmental standards and to upgrade older facilities. With clear environmental objectives and considerations in mind, the goal for the municipal government should be upgrading existing coal facilities by increasing efficiency—rather than adding new coal plants or expanding old ones. Such an approach would also help address pollution issues in Taiwan.

Given that Nuclear Power Plant One (Chinshan) is slated to completely terminate operations in summer 2019, the short-term increase of goal usage would be necessary to fill the gap. In late October 2018, Taiwan's Ministry of Economic Affairs issued a statement strongly defending the Shenao (申澳) Power Plant, saying that the coal-burning plant plays a key role in Taiwan's energy transition[288] The Ministry of Economic Affairs argued that the plant

---

286. Energy Transition Roadmap (能源轉型路徑規劃).
287. Taiwan's Taichung Power Plant Must Reduce Coal Consumption by 24% in 2018," *Taiwan News*, 30 November 2017, https://www.taiwannews.com.tw/en/news/3309637 (accessed 20 May 2019)
288. MOEA Clarifes that Shenao Plant is Important for Energy Transition (經濟部澄清深澳電廠可確保電力穩定)", MOEA(經濟), 11 October 2018,

was important to helping the island reach its 2025 power reserve margin target.[289] Accordingly, a survey conducted by Greenpeace Taiwan and summarized by *Focus Taiwan* found that "80 percent of residents in the three northern cities of Taipei, Keelung and New Taipei had not been aware of the plan" and "75 percent of them would not support it."[290] Given environmental and public health concerns, Premier Lai announced on October 12 2018 to cease the construction of the Shenao plant in consultation with the Ministry of Economic Affairs. The Ministry later released a press release saying that it "has decided to stop construction of the Shenao plant."[291]

Numerous countries across the world have used coal as a transitional fuel in the process of replacing nuclear energy with renewable energy. Germany plans to shut down their coal plants in the next 19 years through a scheme that works on developing and deploying renewable energy in the long term while increasing coal use in the short term.[292] Germany is one of Europe's largest producers and burners of coal, which accounted for 40.3% of net power production in 2017. [293] According to a report in *Foreign Policy*, "Germany's

---

https://www.moea.gov.tw/MNS/populace/news/News.aspx?kind=9&menu_id=22333&news_id=81141 (accessed 20 May 2019)
289. Ibid.
290. Kuan-lin Liu, Wu Hsin-yun, Justin Su, and Chen Chun-hua, "Most northern Taiwan residents opposed to coal-fired power plant: poll," *Focus Taiwan*, 27 March 2018, http://focustaiwan.tw/news/asoc/201803270015.aspx (accessed 10 May 2019)
291. "MOEA Ceases Shenao Construction (經濟部決定停建深澳電廠)", MOEA (經濟部), 12 October 2018, https://www.moeaboe.gov.tw/ECW/populace/news/News.aspx?kind=1&menu_id=41&news_id=16170 (accessed 20 May 2019)
292. Justin Wise, "Germany to Shutter all 84 coal-fired plants to fight climate change," *The Hill*, 27 January 2019 https://thehill.com/policy/energy-environment/427182-germany-to-shut-down-all-84-of-its-coal-fired-plants-to-fight (accessed 20 May 2019)
293. "Germany: Overview," US Energy Information Administration, August 2016, https://www.eia.gov/beta/international/analysis.php?iso=DEU (accessed 20 May 2019);

carbon emissions haven't declined for nearly a decade and the German Environment Agency calculated that Germany emitted 906 million tons of CO2 in 2016."[294] After years of falling behind on emission targets, the country finally announced in early 2019 that it plans to shut down all 84 of its coal-fired power plants within the next two decades to meet its international commitments regarding the fight against climate change.[295] The roadmap also includes $45 billion in spending to mitigate for losses in the transition from coal generation.[296]

Another country that plans to decommission its nuclear power plants and phase out nuclear power with short-term, increased coal usage is Japan. Japan is the only G7 country that still plans to build new coal-fired power stations.[297] According to *Nikkei Asian Review*, "Japanese government officials justify their reliance on coal by citing cost, security of supply concerns, and the need for a diverse energy mix," with a Ministry of Economy, Trade and Industry (METI) representative noting that "[coal] is cheap and more economical with scale."[298] Yet, due to international criticism, Japan has been developing a path to decrease its reliance on coal. Furthermore, the Japanese Ministry of the Environment told *The*

"Net public electricity generation in Germany in 2019," Fraunhofer ISE May 2019, https://www.energy-charts.de/energy_pie.htm (accessed 19 May 2019)
294. Paul Hockenos, "Germany Is a Coal-Burning, Gas-Guzzling Climate Change Hypocrite," *Foreign Policy*, 13 November 2017, https://foreignpolicy.com/2017/11/13/germany-is-a-coal-burning-gas-guzzling-climate-change-hypocrite/ (accessed 20 May 2019)
295. "Germany to Shutter all 84 coal-fired plants to fight climate change"; "Germany to phase out coal by 2038 in move away from fossil fuels," *CNBC News,* 26 January 2019, https://www.cnbc.com/2019/01/26/germany-to-phase-out-coal-by-2038-in-move-away-from-fossil-fuels.html (accessed 19 May 2019)
296. "Germany to phase out coal by 2038 in move away from fossil fuels"
297. "Japan: Second Worst in G7 at Reforming Fossil Fuel Subsidies," NDRC, 5 June 2018, https://www.nrdc.org/experts/han-chen/japan-second-worst-g7-reforming-fossil-fuel-subsidies (accessed 20 May 2019)
298. Eri Sugiura and Akane Okutsu, "Why Japan finds coal hard to quit," *Nikkei Asian Review,* 21 November 2018, https://asia.nikkei.com/Spotlight/Cover-Story/Why-Japan-finds-coal-hard-to-quit (accessed 19 May 2019)

Asahi Shimbun that in principle it will not penalize the addition of new coal-fired plants or boilers to existing facilities and it will coordinate with METI following environmental impact assessments before finalizing construction decisions.[299]

Taiwan has already demonstrated its commitment as a global partner in counteracting climate change—even though it is unable to sign the Paris Agreement, the Taiwan government passed the Greenhouse Gas Reduction and Management Act, which sets five-year, carbon reduction targets.[300] Though in the short-term Taiwan is unable to decrease coal-usage, it will be crucial for its government to lay out a roadmap for cutting the island's reliance on coal in the long-run, by making plans to shutter coal-fired power plants, improve emissions, or reduce electricity usage.

Reform the Electricity Sector and Ensure Strong Oversight

In 2017, Taiwan passed its "Electricity Act [電業法]" to allow other companies—aside from Taipower—to directly buy renewable energy and decrease their carbon footprints, foster competition in electricity generation and retail, and encourage renewable energy financing.[301] Although this was a major breakthrough for the island's electricity sector, transmission and distribution of electricity

---

299. "Japan to rule out coal-fired plants as international criticism rises," *The Asahi Shimbun,* 28 March 2019, http://www.asahi.com/ajw/articles/AJ201903280066.html (accessed 19 May 2019)
300. "First Year Implementation of the Greenhouse Gas Reduction and Management Act," Taiwan EPA, 24 August 2017, https://www.epa.gov.tw/eng/F7AB26007B8FE8DF/c36f5c8a-5448-4ae6-a4a8-aa71cb76abae (accessed 20 May 2019)
301. "Taiwan Passes Law to Liberalize Green Energy Supply," *Focus Taiwan*, 11 January 2017, http://focustaiwan.tw/news/aeco/201701110029.aspx (accessed 20 May 2019)

remained state-owned, under the reins of Taipower. Furthermore, given that Taipower still has control over transmissions lines, the utility company is unlikely to be incentivized to promote the development of renewable energy by other companies and instead will likely seek to maintain dominance in its traditional areas of business in coal and nuclear power. In such circumstances when the infrastructure is not completely independent, green investors would bear the risks that the physical energy infrastructure may not integrate smoothly with any new players. Political impetus thus would be crucial in compelling Taipower to separate generation and transmission in order to attract private investment down the line. At the same time, the government should take the opportunity to invest in basic infrastructure, such as smart grid and meter, to enable long-term electricity reform.

For example, Japan encountered an energy crisis in the aftermath of Fukushima, but took the opportunity to reform its electricity sector. Prime Minister Shinzo Abe has released numerous plans towards deregulation of Japan's electricity sector.[302] The first step is to set up an independent system operator (ISO), alongside a cross-regional coordination of transmission operators (OCCTO), to facilitate the process by 2015. The ISO will manage the flow of electricity with Japan's national grid and help with deregulation, while the OCCTO will maintain the supply-demand balance for transmission and distribution networks.[303] The next step will focus on deregulation in the retail electricity market by 2018, opening it up for competition, and will be followed by the final step which aims to separate

302. "Reforming Japan's Electricity Sector: Abe's Push for Deregulation, Interview with Koichiro Ito," *National Bureau of Asian Research*, 23 October 2013, https://www.nbr.org/publication/reforming-japans-electricity-sector/ (accessed 19 May 2019)

303. "Reforming Japan's Electricity Sector: Abe's Push for Deregulation, Interview with Koichiro Ito"; Peter Weigand and Sumitaka Matsumoto, "Japan's New Electricity Market," *Electric Light and Power*, 16 July 2014, https://www.elp.com/articles/powergrid_international/print/volume-19/issue-7/features/japan-s-new-electricity-market.html (accessed 20 May 2019)

transmission from regional monopolies by 2020 and allow greater access to transmission lines.[304] Such an approach has allowed Japan to experience positive changes, allowing new participants within its electricity market and a gradual reduction in electricity prices. However, Japanese scholars have noted that while market liberalization allows for falling prices in the short-term, rates may increase in the long-term with the introduction of renewable energy.[305]

Japan was once like Taiwan, with its electricity sector held by a monopoly. However, following the Fukushima incident, the country decided to implement radical changes and allow for electricity reform. These are difficult but valuable and important steps that Taiwan could consider replicating or adapting to gradually deregulate its electricity sector and allow for market-driven competition. Accordingly, in the long run, customers would have more utility providers to choose from and lower energy prices, while the government could collaborate with the private sector to allow for further innovation to optimize the grid and connectivity—a win-win for both sides. In short, Taipower needs to set in motion moves toward deregulation.

Provide Higher Incentives for Renewable Energy

With the goal of achieving 20% renewables by 2025—and solar PV constituting 73% of that share—regulations on solar PV would be key in meeting the renewables target.[306] Taiwan has relatively high

---

304. "Reforming Japan's Electricity Sector: Abe's Push for Deregulation, Interview with Koichiro Ito"
305. "Japan's Electricity Shake-up Gives Power to the People," *The Japan Times*, 31 March 2016, https://www.japantimes.co.jp/news/2016/03/31/reference/japans-electricity-shake-gives-power-people/#.XONAcy2ZO9Y (accessed 20 May 2019)
306. Energy Transition Roadmap (能源轉型路徑規劃).

solar potential, particularly in the southern part of the island, and, as such, solar power is a potential low-hanging fruit for Taiwan's energy transformation. However, solar panels are not being installed at a fast-enough rate and are falling behind schedule.[307] The challenges to this are found in limited land availability and a lack of support from local governments in more remote areas, some of which simply do not have access to grid or transmission lines.[308] The government has drawn from international practices and adopted a Feed-in Tariffs (FIT) policy that guarantees electricity purchase prices, guaranteed access to the gird, long-term promotion of renewables, and future price updates through tariff notices.[309] However, Taiwan's local content requirements in its investment regulations require firms to source materials and supplies from the one and only feeder provider – Taipower – in order to be eligible to receive benefits of the PV policy. Taipower should increase transparency in its rule-making process and release greater guidance on means of attracting foreign investors.

Taiwan's FIT scheme was introduced in 2010 and PV installations have been working relatively well due to FIT pricing. However, the government's wholesale purchase price is dropping each year—giving companies and potential investors less incentive to make down payments and enter the market, while the government is

---

307. "Will Taiwan Meets Its 20GW Solar Goal by 2025," PV Magazine, 7 September 2018, https://www.pv-magazine.com/2018/09/07/will-taiwan-meet-its-20-gw-solar-goal-by-2025/ (accessed 20 May 2019)
308. "Renewable Energy in Taiwan," Lexology, 30 December 2018, https://www.lexology.com/library/detail.aspx?g=95239322-8d6f-47d3-97b5-af49f89a4dc8 (accessed 20 May 2019)
309. "Taiwan Green Industry Report （台灣綠色產業報告）," IEK (台灣工研院), April 2018
https://www.google.com/url?sa=t&rct=j&q=&esrc=s&source=web&cd=7&ved=2ahUKEwiD7Oq-sKviAhXLpFkKHa6wCjoQFjAGegQIBxAC&url=https%3A%2F%2Fwww.greentrade.org.tw%2Fzh-hant%2Fstructure%2Ffiledl%2F37846%2F33955&usg=AOvVaw0D7p3WYQo3YCGIh6oqTe1T (accessed 20 May 2019)

struggling to meet subsidy promises. [310] According to EnergyTrend, "The Taiwanese Ministry of Economic Affairs (MOEA) has announced a 10.17% decrease to next year's feed-in-tariff (FiT) rates for solar PV installations, which is much higher than the average decrease of 4.25% in the global PV industry. This will make 2019 a tough year for Taiwan's PV industry, with wider-than-expected impacts on the whole market."[311] Though companies have planned construction of solar PV nearing 1GW, developers still face construction challenges and concerns over profit.[312] After all, a sudden cut in FIT rates do not help development of the solar industry and Taiwan will need to set a more attractive rate to ensure pricing.

On top of pricing, solar PV fields oftentimes require negotiations over land rights that allow for connection to the feeder, and such processes could take years before operations can convene. As such, engagement with local engagements would play a key role in guaranteeing success in grid connectivity. Further, a few proposed construction sites are in the coastal wetlands of Chiayi, which raises environmental concerns regarding sensitive ecosystems.[313] Due to the aforementioned reasons, Taiwan's solar PV development is behind schedule and the government will need to put forth clear policy guidelines in revisiting whole pricing, as well as managing and mitigating concerns regarding land rights and the environment.

Taiwan's conditions are optimal for offshore wind development. The Taiwan Strait has wind speeds of 12m/s and "guaranteed power

---

310. "Taiwan Green Industry Report (台灣綠色產業報告)"
311. "Obscured policies in Taiwan's FIT scheme to impact on sustainable development of local solar supply chain," EnergyTrend, 17 December 2018, https://m.energytrend.com/research/view/12839.html (accessed 19 May 2019)
312. Ibid.
313. Yeh Tzu-kang and William Yen, "Solar Panel Installation Will Not Hurt Bird Habitat: Chiayi," *Focus Taiwan*, 12 May 2019, http://focustaiwan.tw/news/asoc/201905120011.aspx (accessed 20 May 2019)

prices of TWD 5,849/MWh (US $189/MWh)."[314] This has garnered the attention of numerous international firms and investors, such as Denmark's Ørsted and Germany's Wpd.[315] In addition, the Taiwan government had adopted FIT systems that are traditionally more competitive than European rates in the global wind industry and has likewise provided clear policy guidance and coordinated with local banks on project finance.[316] However, a common concern in wind project investments is time—the long development timeframes of offshore wind infrastructure is a key consideration for developers, which makes any price or tariff adjustments particularly sensitive. According to the Global Offshore Wind Taskforce, "In late 2018, the Taiwanese Government proposed revisions to the feed in tariff (FIT)…a cap on the rate of 3,600 full load hours; and the removal of the optional accelerated FIT for the first ten years…[and] [a]s a result, it is estimated that the revenue reduction will be approximately a third of that previously proposed."[317]

Another challenge is the lack of offshore wind infrastructure in Taiwan. It will take years of effort to build up local supply chains to support major offshore projects. Despite its little experience, the Taiwan government has already put in a large amount of effort in accelerating contract permits, increasing administrative efficiency, and supporting connectivity infrastructure on grid connections as

---

314. Ilaria Valtimora, "Offshore Wind: Is The Future In Taiwan At Risk?", Renewable Energy World, 22 January 2019, https://www.renewableenergyworld.com/articles/2019/01/offshore-wind-is-the-future-in-taiwan-at-risk.html (accessed 19 May 2019)
315. Ibid.
316. Stine Jacobsen,"Offshore Wind Power Firms See Taiwan as a Battleground to Expand in Asia," *Reuters*, 30 April 2018, https://www.reuters.com/article/us-taiwan-windpower/offshore-wind-power-firms-see-taiwan-as-a-battleground-to-expand-in-asia-idUSKBN1I11IV (accessed 20 May 2019)
317. Alastair Dutton, "Challenges lay ahead for Taiwan," *Wind Power Monthly*, 4 February 2019, https://www.windpowermonthly.com/article/1524579/challenges-lay-ahead-taiwan (accessed 19 May 2019)

soon as possible.[318] In April 2019, according to news sources, "The EECT [European Chamber of Commerce in Taiwan] estimated that if the target of 5.5 GW is reached, Taiwan could receive NT $880 billion (US $28.4 billion) in investment in the sector, [creating] 20,000 extra jobs, additional construction projects and the development of a whole new domestic supply chain for the industry."[319]

Lastly, an important aspect to bring to the picture is natural disasters. Taiwan is located on the highly active, earthquake-prone Pacific Ring of Fire and is also within the path of summer typhoons. To successfully build offshore wind turbines in an earthquake zone, the turbines must be able to withstand pressures from strong winds and thus be constructed with sturdy enough materials that ensure both operational excellence and safety of the invested capital. This requires technological innovation, as well as investment, from the government and private investors. Additionally, the government would need to assess environmental considerations, especially in the western wetlands and conservation sites off the coast. Consultations with fishermen and other affected groups should be convened and standards should be put in place in regards to compensation to these individuals in the construction of offshore wind farms in order to fully strike a balance between energy security, social safeguards, and environmental preservation.

---

318. Lee, Tsai & Partners, "Renewable Energy in Taiwan," Lexology, 3 April 2019, https://www.lexology.com/library/detail.aspx?g=d78ac5fd-cd61-4c49-b395-8fb2ea5da35d (accessed 20 May 2019); "Taiwan Offshore Wind Farm Projects: Guiding Investors through the Legal and Regulatory Framework," Jones Day, February 2018, https://www.jonesday.com/files/Publication/391679ef-9fca-4142-a4c1-df4e06b1d574/Presentation/PublicationAttachment/ebe8c0fb-1e42-4d95-be45-f914d4a6f1f9/Taiwan%20Offshore%20Wind%20Farm%20Projects%20White%20Paper.pdf (accessed 20 May 2019)
319. Matthew Strong, "Taiwan wind energy could result in 20,000 new jobs in 2025," *Taiwan News*, 25 April 2019, https://www.taiwannews.com.tw/en/news/3687974 (accessed 19 May 2019)

Taiwan has great potential and solid policies already in place that can lead to the long-term development and sustainability of offshore wind power. The government has made progress in attracting foreign companies and this is a positive start—however, policymakers must continue to ensure environmental conservation, maintain competitive prices, and address global industry concerns on ladder tariffs.

Conclusion

All in all, Taiwan's energy future is a complex issue that includes national- and economic-security considerations and is beset by a relatively inflexible bureaucracy. The price for not moving forward and continuing down the path of a nuclear phase-out is too high—putting the island's population of 23 million at risk and setting a stark example for neighboring countries. The price of not changing existing energy policies writ large—even irrespective of the nuclear debate—is likewise high, given low energy reserves and an economy driven by industries with high electricity usage.

Optimistically, leadership on the island under the current Tsai administration appears steadfast in its ambitious goal of phasing out nuclear power, despite the referendum outcome. Positive momentum has been made in that direction, with Nuclear Power Plant One set to be fully decommissioned in 25 years, with no moves to renew the license of its final unit once it expires in July 2019.[320]

However, tough decisions backed by sufficient political will are needed if such a goal is to achieved. A clear roadmap should be developed to build up gas capacity, increase coal usage in the short-

---

320. Keoni Everington, "EPA approves decommissioning plan for Taiwan's No. 1 nuke reactor," *Taiwan News*, 16 May 2019, https://www.taiwannews.com.tw/en/news/3702875 (accessed 1 June 2019)

term, and put direct investment and resources into expanding Taiwan's renewable energy sector. In addition, Taipower will need to shoulder the responsibility of ensuring oversight in infrastructure investments and continue conducting public consultations with key stakeholders and other vested parties to ensure grid connectivity and sustainable development.

Finally, and perhaps most critically, is the reality that Taiwan's energy future and the future of nuclear energy on the island is intrinsically tied to partisan shifts. The Tsai administration and any potential future DPP administration is likely to remain committed to phasing out nuclear power or, at the very least, set in motion plans and policies conducive to such a goal. However, the island holds elections every two years—alternating between local and national—and while the latter determines which party leads the Executive and Legislative branches, and by extension, decision- and policymaking on energy issues, the former serves as a bellwether for party performance and local election results may compel a recalibration or change of pace in policies to better match the mood of the electorate. In short, while options may be available to end nuclear power, the true test will lay in whether there's sufficient political support from Taiwan's people to bear the short-term costs for the sake of long term sustainability.

Economic Considerations of Nuclear Power Deployment in Saudi Arabia

Ali Ahmad

Introduction

Saudi Arabia's plans to acquire nuclear power are ambiguous. In 2012, the Kingdom announced its intention to build 16 nuclear reactors by 2032, which was later delayed to 2040.[321] More recently, the "Saudi Vision 2030", championed by the now Crown Prince, Mohamad Bin Salman, did not explicitly mention nuclear power as an option for the Kingdom, though it was promptly followed by the "National Transformation Program 2020", which included the need to identify and the prepare the construction location of Saudi's first nuclear power plant, localize parts of the nuclear fuel cycle.[322] In September 2018, Saudi Arabia short-listed Westinghouse USA, Russia's Rosatom, South Korea's KEPCO, and China's CNNC for its two large-scale reactors. However, there is no financing, contractual or technical model envisioned yet. On the SMR's front, the Kingdom is in advanced stage with Korea's KAERI to perform feasibility studies on at least two SMART reactors of 100 MWe generation capacity for desalination.

321. WNA. (2017). Nuclear Power in Saudi Arabia. Retrieved Oct 2, 2017, from http://www.world-nuclear.org/information-library/country-profiles/countries-o s/saudi-arabia.aspx

322. Saudi National Transformation Program. (2016). Retrieved Otc 2, 2017, from http://vision2030.gov.sa/en/ntp

Saudi officials offer multiple motivations to construct nuclear power plants in the Kingdom. Perhaps the most commonly articulated one is a desire to get away from the near-complete reliance on hydrocarbon resources to produce electricity and desalinated water. Such reliance is seen to lead to the depletion of national oil and gas reserves but is also perceived to have an opportunity cost associated with forgone oil export revenues. Others suggest that interest in nuclear power is also a response to Iran's acquisition of nuclear technology.[323] Whatever the motivations behind nuclear power might be, the acquisition of nuclear technology will entail major economic and non-economic costs.

This paper examines the main economic considerations for building nuclear power plants in Saudi Arabia. It focuses on issues such as large vis-à-vis small reactor options, cost of electricity generation, localization potential, and the role of renewables, particularly solar power. The paper also discusses the measures required and investments needed to achieve high penetration of renewables in the Kingdom.

The Electricity Sector in Saudi Arabia

The Kingdom's power generation capacity is built on conventional thermal plants fueled by a mix of crude oil, heavy fuel oil, natural gas and other petro-residuals, as shown in Figure 1.[324] All of the natural gas produced within the Kingdom is consumed domestically while electricity generation is dominated by gas turbines which offer

---

323. Henry Sokolski. "In the Middle East, soon everyone will want the bomb." Foreign Policy. May 2018.

324. It should be noted that Saudi Arabia's oil share in the power sector is higher than most other countries, resulting in significant opportunity cost, particularly when oil prices are high.

a cheap way of meeting demand.[325 326] The Saudi Electric Company (SEC), a vertically integrated electricity company, controls 71% of the generation capacity and is responsible for transmission and distribution across the country.[327] More recently, government reforms are pushing for unbundling of the generation, transmission and distribution sectors in order to encourage the entry of new producers thereby increasing efficiency and decreasing government spending.[328]

Figure 1: Electricity generation in Saudi Arabia by type of fuel and generation technology (ST= steam turbine, CCGT= Combined Cycle Gas Turbine, GT=Gas Turbine, Diesel Generators)

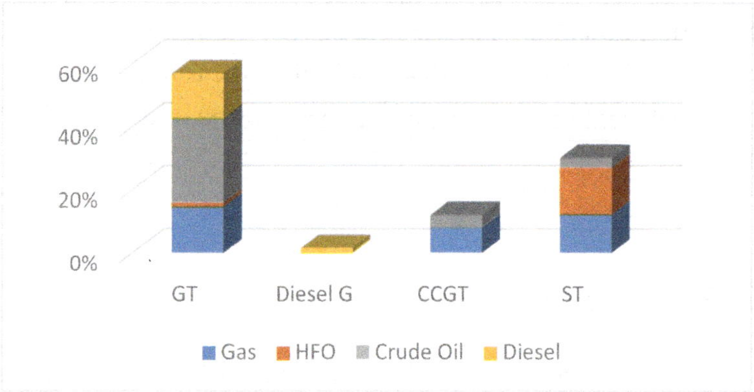

325. Segar, Christopher. "Getting it Right : Renewables Augment Gas Saudi Energy Mix." *The Journal of the International Energy Agency - EIA Energy,* issue 7 - 4th quarter (2014): 40.
326. Rabia Ferroukhi, Arslan Khalid, Diala Hawila, Divyam Nagpal (IRENA), Laura El-Katiri, Vasilis Fthenakis and. *Renewable Energy Market Analysis: The GCC Region,* IRENA, 2016.
327. Nachet and Aoun, The Saudi electricity sector: pressing issues and challenge.
328. Ibid.

By 2032, power generation capacity in the Kingdom is expected to reach 120 GW.[329] The increase in electricity demand will be mostly driven by population growth, a fast-growing economy and an increase in consumption fueled by cheap energy prices and high government subsidies.[330]

Based on 2014 data, the maximum peak load reached in Saudi Arabia was approximately 56.5 GW occurring during week 36 (September 1-7, 2014), while the minimum peak reached was 31.88 GW occurring during the first week of that same year (December 30, 2013-January 5, 2014). The high peak load occurred in summer and low peak load occurred in winter. The difference between highest and lowest load in summer was 8.77 GW, and the difference in the winter was 7.55 GW.[331] The summer pattern is specific to a number of countries including the Gulf countries where air conditioning is widely used in summer and barely needed for heating in winter. See Figure 2.

---

329. Saudi Arabia *Annual Statistical Booklet for Electricity and Seawater Desalination Industrie, 2016*
330. CIA, *The World Factbook 2016-17.*
331. Philippe Chite and Ali Ahmad, "Requirements for High Solar Penetration in Electricity Production in Saudi Arabia", 2017, Energy Transitions in the GCC, Gulf Research Center, forthcoming.

Figure 2: Hourly load in Saudi Arabia during summer (left) and winter (right) – Source: Chite and Ahmad, 2017

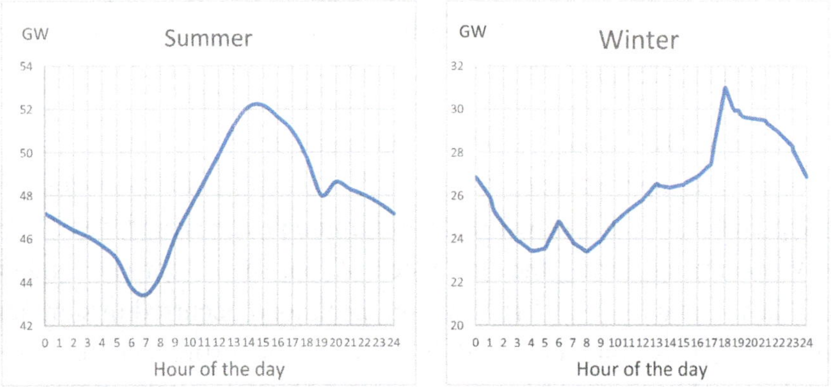

Nuclear Technology Options

There are multiple reactor technologies that are available for deployment in Saudi Arabia. Many of the reactors that have already been sold and are being operated or under construction have relatively large power capacities (Table 1).[332] More detailed technical descriptions of reactor technologies and their relevance to the Middle East market can be found in the literature.[333] In the Gulf Cooperation Council (GCC), the South Korean APR-1400 technology is the only one currently under construction in the United Arab Emirates.[334]

332. There are numerous other designs that are potentially available for sale but these are not listed since they have so far not been constructed or under construction.
333. Mohamed S. El-Genk, "On the Introduction of Nuclear Power in Middle East Countries: Promise, Strategies, Vision and Challenges," *Energy Conversion and Management* 49 (2008): 2618–28.
334. WNA. (2017). Nuclear Power in UAE. Retrieved Oct 2, 2017, from http://www.world-nuclear.org/information-library/country-profiles/countries-t-z/united-arab-emirates.aspx

Table 1: Current reactor designs available for Saudi Arabia

Country	Technology	Reactor Design	Capacity (MWe)
South Korea	PWR	APR-1400	1400
France	PWR	EPR	1600-1700
USA/Japan	PWR	AP1000	1110
Russia	PWR	VVER	1000
USA/Japan	BWR	ABWR	1350
India/Canada	PHWR	PHWR/ACR-700	700

Saudi Arabia's ambitious plans for nuclear power has attracted a number of nuclear vendors. In 2013 both GE Hitachi Nuclear Energy and Toshiba/Westinghouse signed contracts with Exelon Nuclear Partners (ENP), a division of Exelon Generation, to pursue reactor construction deals with KA-CARE.[335] The reactor designs proposed include the ABWR, the ESBWR and the AP1000. The French companies Areva and EDF have also been aggressively moving into Saudi Arabia. In 2015, KA-CARE signed nuclear cooperation agreement with France to study the feasibility of constructing two EPR reactors while providing training on safety and waste disposal. Saudi Arabia and China signed a MoU in 2016 to build of High-temperature gas-cooled reactors (HTGRs) and in 2017 to study the feasibility of building HTGRs.[336]

---

335. WNN, "Teaming up for Saudi Bids," *World Nuclear News*, September 9, 2013, http://www.world-nuclear-news.org/NN-Teaming_up_for_Saudi_bids-0909137.html.
336. WNA. (2017). Nuclear Power in Saudi Arabia. Retrieved Oct 2, 2017, from http://www.world-nuclear.org/information-library/country-profiles/countries-o-s/saudi-arabia.aspx

More recently, Saudi Arabia sent a "request for information" to nuclear reactor vendors around the world, a step perceived as a first step towards opening a formal tender.[337] As a response, Westinghouse, which promotes the AP1000 design, has been reported to be in discussion with US-based companies to forma bidding consortium.[338] With reports of Russian and South Korean companies also planning to bid, nuclear vendors worldwide look at Saudi Arabia as one of the most promising markets, especially with if KA-CARE's plan to build 16 reactors is materialized.

Small Modular Reactors

Saudi Arabia's interest in small modular reactors seems to be serious. In March 2015, the Kingdom signed an MoU with South Korea to conduct a three-year study to review the feasibility of constructing SMART reactors in Saudi Arabia. The agreement also calls for the two countries to cooperate on the commercialization and promotion of the SMART reactor to third countries.[339]

A number of SMR developers have argued that there are multiple motivations to pursue smaller designs, directed both at large industrialized countries and developing countries. One motivation is the high upfront capital cost of standard reactors, which is beyond the financing capacities of many utilities and countries. Another is to expand nuclear power to countries with relatively small electrical-grid capacities; a gigawatt-scale reactor could destabilize a small

337. Reuters, "Westinghouse discussing group bid for Saudi nuclear tender", Nov 21, 2017, from https://www.reuters.com/article/us-saudi-nuclear-usa-exclusive/exclusive-westinghouse-discussing-group-bid-for-saudi-nuclear-tender-sources-idUSKBN1DL1BF
338. Ibid.
339 WNN, "Saudi Arabia team up with South Korea on SMART", *World Nuclear News*, March 04, 2015, http://www.world-nuclear-news.org/NN-Saudi-Arabia-teams-up-with-Korea-on-SMART-0403154.html.

grid. What is interesting about the SMART partnership between Saudi Arabia and South Korea is that the Kingdom falls outside the "niche" SMR market, as advocated by SMR vendors.

However, other factors that have been offered as motivations for SMRs are claims to potentially greater safety due to the reliance on passive features and enhancement in public acceptability. In the case of the Saudi SMART venture, the technology is also capable of producing desalinated water besides generation electricity.[340]

The downside of SMRs deployment in Saudi Arabia, however, is the loss of economies of scale – smaller nuclear reactors are typically more expensive on a per unit cost basis. Detailed and carefully conducted elicitations showed that even experts drawn from, or closely associated with, the nuclear industry expect SMRs to cost more per kW of capacity than currently operating reactors.[341] Based on the average of expert estimates of the extra per kW cost for SMRs, the percentage increase expected ranged from 12% for 225 MWe reactors to a whopping 120% for 45 MWe reactors.

There are currently dozens of SMR designs under development. Some of these are still in the conceptual design phase, many are still in the R&D phase and only four have been licensed or currently are under construction.[342] The details of these SMRs are summarized in

---

340. WNN, "Saudi Arabia and Korea further SMART cooperation", *World Nuclear News*, September 03, 2015, http://www.world-nuclear-news.org/NN-Saudi-Arabia-and-Korea-further-SMART-cooperation-03091501.html
341. Ahmed Abdulla, Inês Lima Azevedo, and M. Granger Morgan, "Expert Assessments of the Cost of Light Water Small Modular Reactors," *Proceedings of the National Academy of Sciences* 110, no. 24 (2013): 9686–91.
342. Hadid Subki, "Global Trends, Prospects and Challenges for Innovative SMRs Deployment" (presented at the Long-term Prospects for Nuclear Energy in the Post-Fukushima Era, Seoul (Korea), August 29, 2012), http://www.iaea.org/INPRO/5th_Dialogue_Forum/Wednesday,_29.08.2012/Session_IV_Nuclear_Safety_and_Innovation/2._Hadid_Subki_IAEA_0829_no_distribute.pdf

Table 2. Thus far, there is no sign that any of the five remaining GCC countries are considering any of these reactors.

Table 2: Small modular reactors currently under construction or licensed

Country of Origin	South Korea	Russia	China	Argentina
Reactor Technology	PWR	PWR	HTGR	PWR
Status	Licensed	Under Construction	Under Construction	Under Construction
Electrical Power (MWe)	90-100	35	2X105	27

The HTGR Option

Saudi Arabia's interest in SMRs extends to the Chinese high temperature gas reactor (HTGR). In March 2017, KA-CARE and China Nuclear Engineering Group (CNEC) signed an agreement to conduct a feasibility study that will consider "the development of system solutions for the investment and construction of HTGRs."[343] The agreement will also examine cooperation in "intellectual property and the development of a domestic industrial supply chain for HTGRs built in Saudi Arabia."

Since HTGRs operate at a higher temperature compared to water-cooled reactors, they can be used to generate process heat and

343. WNN, "China, Saudi Arabia begin HTGR feasibility study", *World Nuclear News*, May 17, 2015, http://www.world-nuclear-news.org/NN-China-Saudi-Arabia-begin-HTGR-feasibility-study-1705174.html. In cooler countries, HTGR's excess heat can be used for district heating for residential and commercial customers.

hydrogen production as well as electricity.[344] On paper, Saudi Arabia could deploy HTGRs in their industrial regions such as petrochemical compounds and heavy oil recovery systems. Proponents of HTGRs claim that they possess high safety standards.[345] However, based on a review of the operational history of HTGRs, they face some serious technical challenges and are prone to a variety of small failures such as air and oil ingress, which could trigger accidents with severe consequences.[346] [347] Given the importance of the oil and petrochemical sector for the Saudi economy, the consequences of any accident are likely to be immense. Therefore, the Kingdom will be taking on a substantial risk for a return that can be achieved using other sources of energy and technologies.

Cost of Electricity Generation

The conventional way to compare the cost of electricity generated by different sources is to calculate the levelized cost of electricity (LCOE), which can be understood as the ratio of the total cost to the benefits (in this case the electricity produced) with all figures being discounted to the same baseline year. This follows from the standard discounted cash flow methodology, which accounts for the time-

---

344. In cooler countries, HTGR's excess heat can be used for district heating for residential and commercial customers.
345. Zhang, Zuoyi, Zongxin Wu, Dazhong Wang, Yuanhui Xu, Yuliang Sun, Fu Li, and Yujie Dong. 2009. "Current Status and Technical Description of Chinese 2 × 250 MWth HTR-PM Demonstration Plant." Nuclear Engineering and Design 239 (7): 1212–1219
346. M.V. Ramana, "The checkered operational history of high-temperature gas-cooled reactors", 2016, Bulletin of the Atomic Scientists, 72, No. 3, 171–179
347. Matthias Englert, Friederike Frieß, and M.V. Ramana, "*Accident Scenarios Involving Pebble Bed High Temperature Reactors*", 2017, Science & Global Security, 25, 42--55

value of money. This methodology is used to calculate the life cycle cost of producing electricity.

The calculated LCOE for different energy sources are all busbar costs delivered to the grid; i.e., they take into account auxiliary or in-plant consumption of electricity but do not include transmission and distribution costs. Any large-scale expansion of nuclear power in Saudi Arabia would require an expansion of transmission infrastructure. Such costs are not included here, even though it could be significant.

The cost of electricity generated as well as water produced by desalination from any technology depends on a number of parameters. An important factor is the discount rate. For the case of Saudi Arabia, the chosen discount rate is 5 percent. Note that this is a real discount rate, and inflation is implicitly taken into account. This choice may seem somewhat low, but many studies do indeed adopt discount rates of around 5 to 5.5 percent in their evaluations of electricity economics in the GCC.[348]

---

348. Yasser Al-Saleh, Paul Upham, and Khaleel Malik, *Renewable Energy Scenarios for the Kingdom of Saudi Arabia*, Working Paper (Norwich, U.K.: Tyndall Centre for Climate Change Research, October 2008); Center for Engineering Research, *Updated Generation Planning for the Saudi Electricity Sector* (Riyadh, Saudi Arabhi: Electricity & Cogeneration Regulatory Authority, March 2006).

Table 3: Cost of electricity generation in Saudi Arabia of different technologies using subsidized prices

[349]Fuel/ Technology	Nuclear	Gas-light oil / GT	Gas-light oil/ CCGT	Heavy Oil/ ST	Diesel/ Diesel
Fixed Cost ($/MWh)	68.5	10.37	17.35	14.77	66.98
Var. Cost(US$/MWh)	12.25	37.69	8.70	8.23	21.50
LCOE (cent/KWh)	8.08	4.81	2.60	2.30	8.85

Estimates of the cost of electricity generation of various sources in Saudi Arabia are shown in Table 3. These estimates were based on a recent study by Chite and Ahmad.[350] Estimating the cost of generation using prices of subsidized fuel, as usually entailed from Saudi electricity operators, results in very low generation costs where nuclear (and renewables) cannot compete.

A special case to be considered is the cost of nuclear electricity generated by SMRs. As discussed earlier, SMRs are expected to have higher capital costs per kW by somewhere between 12 and 120 percent.[351] In this analysis, SMRs' capital costs are assumed 25 percent higher than for current nuclear reactors, i.e., approximately $7,430/kW.[352] SMRs would also have higher fueling cost because of

350. Philippe Chite and Ali Ahmad, "Requirements for High Solar Penetration in Electricity Production in Saudi Arabia", 2017, Energy Transitions in the GCC, Gulf Research Center, forthcoming.
351. Abdulla, Azevedo, and Morgan, "Expert Assessments of the Cost of Light Water Small Modular Reactors."
352. This cost value is based on the EIA's reported cost for advanced nuclear, which currently stands at 5945 $/kW.

higher uranium requirements.[353] It is also not clear what sets of conditions they would be licensed under,[354] and depending on the safety and security requirements imposed by regulatory authorities, SMRs could have higher fixed and variable operations & maintenance costs. In contrast to these higher costs, the construction time for SMRs is expected to be shorter.[355] For simplicity, it is assumed that all of the cost variables for current (large) LWRs are the same as SMRs with the exception of the capital cost. Then, with these assumptions, the levelized cost of nuclear power from reactors rises from about $81/MWh if large reactors are constructed to over $94/MWh for SMRs.

Baseload Generation: Nuclear versus Natural Gas

Unlike renewables that have zero fueling costs, the cost of natural gas is an important, perhaps overwhelming, component of the cost of generating electricity in a natural gas plant. Therefore, the cost comparison between nuclear and natural gas could vary from country to country. Natural gas prices in countries that use domestic reserves, such as Saudi Arabia, would be significantly lower than countries that import natural gas at international prices, and this would affect the relative economics of power from nuclear reactors and natural gas plants.

The economic competitiveness of gas-fired power plants decreases as the prices of natural gas increase. Ahmad and Ramana estimated

353. Alexander Glaser, Laura Berzak Hopkins, and M.V. Ramana, "Resource Requirements and Proliferation Risks Associated with Small Modular Reactors," *Nuclear Technology* 184 (2013): 121–29.
354. M.V. Ramana, Laura Berzak Hopkins, and Alexander Glaser, "Licensing Small Modular Reactors," *Energy* 61 (2013): 555–64, doi:10.1016/j.energy.2013.09.010.
355. Abdulla, Azevedo, and Morgan, "Expert Assessments of the Cost of Light Water Small Modular Reactors."

the cross-over value between nuclear and natural gas generated electricity for Saudi Arabia is at a natural gas price of \$13.6/mmBTU.[356]

In other words, nuclear is less economical if natural gas prices are lower than the cross-over value. The economic prospects for nuclear power in the Kingdom are not favorable in comparison with natural gas, even if the currently low domestic natural gas prices in Saudi Arabia were to rise substantially.

What if Saudi Arabia were to increase its domestic output and start exporting gas instead of using it in natural gas plants? This is related to what is often termed the opportunity cost. The first thing to note is that such exports will most likely be in a liquefied (LNG) form. Consequently, the costs associated with building infrastructure, liquefaction and shipping should be taken into account. A study on the future of natural gas conducted by the Massachusetts Institute of Technology estimates the cost of liquefaction at \$2.15/mmBTU, shipping of LNG at \$1.25/mmBTU, and regasification at \$0.7/mmBTU.[357] The total of these costs amount to \$4.1/mmBTU.[358] Because of this additional expenditure, it would make economic sense for Saudi Arabia to build a nuclear reactor in comparison to a natural gas plant only if the price that could be obtained on the international market exceeded \$13.6/mmBTU over the period of the lifetime of the reactor.

---

356. Ali Ahmad and M. V Ramana, "Too Costly to Matter: Economics of Nuclear Power for Saudi Arabia," *Energy*, 69 (2014): 682–694.
357. Ernest J Moniz, Henry D. Jacoby, and Anthony J. M. Meggs, *The Future of Natural Gas* (Cambridge, MA: Massachusetts Institute of Technology, 2011).
358. Additionally, LNG plants are very expensive, with long-lead times and significant potential market risks as more countries ramp up to export LNG or as importing countries decide they can boost local gas production using fracking. Since the LNG plants must operate for long periods of time to recover costs, these risk factors ought to further tilt countries without sufficient LNG capacity already built to favor domestic use in power markets instead.

Potential of Solar Power in Saudi Arabia

Aside from meeting the increasing energy demand, integrating solar power in the Kingdom's current energy mix would be both economically and environmentally advantageous. Saudi Arabia has one of the highest potentials for solar energy in the region where annual solar radiation is around 2,200 kWh/m^2.[359] Moreover, integrating renewables within the energy industry would drive economic diversification, create jobs and facilitate the implementation of climate change policies. Compared to nuclear, pairing solar power plants with domestic gas turbines can also help with load balancing more effectively.

The Kingdom recently equipped 30 metrological stations able to conduct very accurate measurements, including one-minute measurements of Global Horizontal Irradiance (GHI), Diffuse Horizontal Irradiance (DHI), and Direct Normal Irradiance (DNI).[360] GHI ranged from an average daily total of 5,700 Wh/m^2 to the highest 6,700 Wh/m^2 with the higher values found in inland areas and lower values found on the coast. DNI values ranged from an average daily total of over 6,474 Wh/m^2 found on western inland areas to an average daily total closer to 5,510 Wh/m^2 found on eastern areas.[361] The solar resources outlined above are optimal for the performance of two main solar technologies that dominate current and future energy projects in Saudi Arabia: Photovoltaics (PV) and Concentrated Solar Power (CSP).

The hourly load pattern shown in Figure 2 above should offer an advantage for renewables mainly solar, as the high load occurs in

---

359. *Electricity Generation Analyses in an Oil-exporting Country: Transition to Non-Fossil Fuel Based Power Units in Saudi Arabia.* A. Farnoosh, F. Lantz, & J. Percebois, *Energy* 69 (2014): 299–308
360. *Renewable Resource Atlas*, K.A.-CARE King Abdullah City for Atomc and Renewable Energy, [https://rratlas.kacare.gov.sa/RRMMPublicPortal/.
361. Erica, Zell et al. "Assessment of Solar Radiation Resources in Saudi Arabia" *Solar Energy* 119 (Elsevier, 2015): 422-438.

the summer, in step with the maximum output of solar PV or CSP systems. However, PV systems generally suffer from reduced power output during the summer due to high ambient temperature affecting the performance of solar cells. On the other hand, CSP systems are less affected by high ambient temperature but are more sensitive to weather conditions like haze or sandstorms which lower the performance of solar CSP plants.[362]

In terms of costs, PV costs are divided into module costs (direct cost of photovoltaic modules); "hard costs" (inverter, racking, electrical equipment, etc.); and "soft costs" (labor, permitting fees, etc.). Module prices have followed a learning rate of 20 percent over the long term for the last 10 years.[363] Non-module prices, which are also known as BOS (Balance of System) are also decreasing, nearly at the same rate. Future improvements in the PV technology cost should come from a combination of improving power electronics, reducing supply chain complexity and cost, and decreasing installation costs and margins as markets mature.[364]

The dramatic decline in solar PV costs is particularly relevant because of the long construction period of nuclear projects, a decade at the very least.[365] Even assuming that a nuclear power project is given the go-ahead in 2018, it will likely be 2028 by the time it starts generating electricity. On the other hand, there is little reason to

---

362. *CSP Prospects in Saudia Arabia* , www.csptoday.com/menasol2014, 2014.
363. http://trade.ec.europa.eu/doclib/press/index.cfm?id=1461.
364. *Ibid.*
365. According to the IAEA's figures for construction periods, the global weighted average reactor construction time is 96.6 months, or about eight years. See IAEA, *Nuclear Power Reactors in the World: 2013 Edition* (Vienna: International Atomic Energy Agency, 2013). In these figures, the construction period is defined as the time between the "first pouring of concrete to the connection of the unit to the grid". However, some costs that are incurred before the first pour of concrete—for example, money spent on ordering components that take a long time to manufacture. If one were to include those as well, there would be cash flows from a project initiation for about ten years.

expect the costs of nuclear power to decline substantially. Historically, costs of nuclear power have only increased.[366]

Recent solar PV projects benefit from lower prices. For example, the Dewa project in Dubai rated 260 MW is priced at $328 million, giving a capital cost of 1,225 $/kW. The plant is expected to produce electricity at world record of 5.85 cent/kWh. Prices are heavily influenced by the project location. In Saudi Arabia, these costs should be less than those in the U.S. and Europe. Costs should decrease even further as more PV projects are installed and some parts of the PV system are manufactured locally.

Investments Needs

The total investment for the 2012 KA-CARE plan is expected to reach about $360 billion by 2030, with CSP technologies representing the highest costs (see Figure 3).[367] A cheaper scenario involves adding gas turbines and replacing nuclear reactors with CCGT power plants, which are cheaper and have a shorter construction time. In that case, the total investments needed are approximately $150 billion, nearly half the investments needed by 2012 KA-CARE plan (see Figure 4).

---

366. Hultman and Koomey, "The Risk of Surprise in Energy Technology Costs"; Arnulf Grubler, "The Costs of the French Nuclear Scale-up: A Case of Negative Learning by Doing," *Energy Policy* 38, no. 9 (2010): 5174–88.
367. Calculations used the following capital costs: 2,500 $/ KW for PV, 6,800 $/KW for CSP with six hours storage and $5,530 for nuclear reactor Olson, Schlag, Patel, and Kwok, *Capital Cost Review of Power Generation Technologies;* Ali Ahmad, and M.V. Ramana, Too Costly to Matter: Economics of Nuclear Power for Saudi Arabia. *Energy* 69 (2014): 682-694

Figure 3: Investment in Billion US$ needed to implement the original KA-CARE plan (16 GW PV, 25 CSP and 17.6 GW Nuclear)

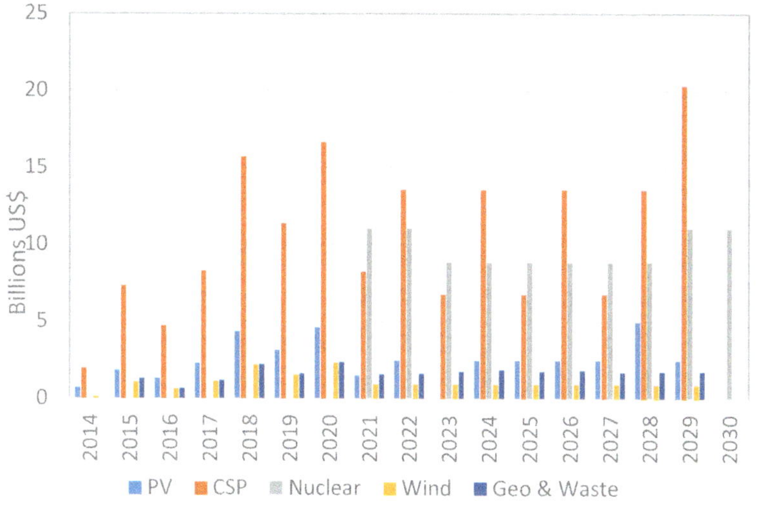

Figure 4: Investment in billion $ needed to implement the modified plan (16 GW PV, 10 CSP and 17.6 GW CGGT and 30 GW GT)

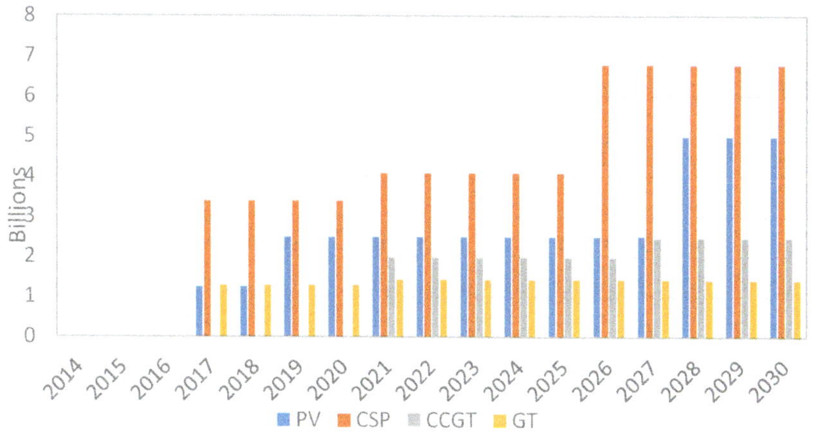

High solar penetration in the Saudi electricity sector will depend the support of the political leadership, the type of technology, its capacity and associated costs, the energy market structure, the investment climate, and, finally, the permitting rules and regulations.[368] The "Saudi Vision 2030" shows strong interest in renewables among political leaders, but the energy sector is still dominated by oil and remains highly subsidized, which further leads to very low fuel and electricity prices encouraging high consumption.

All the existing entities needed for leading the energy transition are still new or yet to be established; however, it is likely that Saudi Aramco will play an important role in implementing this transition. Saudi Arabia will still need further research on the effect of sandstorms, soiling, aging, and high temperatures on renewable energy systems. The country should also promote energy efficiency and storage schemes, smart grid systems, and interconnection with neighboring countries. Moreover, the localization of renewables should be a top priority for policy makers since the Kingdom can rely on international partnerships for technology transfers that could facilitate its energy transformation.

The Localization Question

Localization has been mentioned as an important theme in the Saudi 2030 vision, and was further emphasized in the National Transformation Plan 2020. In general, countries seek to promote local content requirements to promote public support, support new industries, and add new jobs.

---

368. *Moving towards 100% Renewable Electricity in Europe & North Africa by 2050: Evaluating progress in 2010.* PricewaterhouseCoopers, 2011.

According to the National Transformation Plan 2020, the Saudi government aims to promote local content in the nuclear and renewable energy sector simultaneously.[369] For the nuclear energy sector, the Kingdom plans to introduce nuclear power to the energy mix of the Kingdom with a target of 25% to 30% local content that includes nuclear fuel cycle activities such as in uranium production.[370] To support the legislative efforts, building infrastructure, training human capabilities, localizing renewables technology, SMRs and the nuclear fuel cycle, 5 billion Saudi Rial will be spent by 2020.[371]

Saudi Arabia's intention to localize nuclear industries and the fuel cycle clearly contradicts attempts by SMR vendors to cut costs through, factory-build, serial production models, unless the Kingdom intends to build tens of SMART reactors across its territory, though this does not seem to be a realistic option given the current lack of economic, industrial, and human capacities. Furthermore, the possibility of linking all GCC countries with one electric grid makes the economic case for SMRs in the Kingdom even weaker since such gird interconnections would be ideal to improve the economics of renewables.

On the other hand, the localization potential for renewables, particularly solar, in the Kingdom is high. It is also easier to achieve compared to the complex and political nuclear localization. Additionally, it is not surprising that the decentralized decision-making process involved with planning and constructing renewable energy projects plays an important role in advancing its localization potential.

---

369. National Transformation Program. (2016). Retrieved October 2, 2017, from http://vision2030.gov.sa/en/ntp
370. It is not clear for now whether "uranium production" includes enrichment.
371. Ibid.

Investments in renewable energy in Saudi Arabia should be accompanied by localization of some part of the value chain in the Kingdom. For PV, the localization of BOS (balance of the system) activities can be the first step; this could include manufacturing support structures, trackers, mounting hardware, electric protection devices, wiring, monitoring equipment, and installation. As a second setup in localization the Kingdom can build a manufacturing facility in the country.

The investments needed for localizing a 1GW/yr PV manufacturing facility range between 1 and 2 billion US$ per plant. By 2030, a 7 GW PV plant should create around 21,000-50,000 direct jobs and 70,000-140,000 indirect jobs in the Kingdom.[372]

Conclusion

Nuclear power has been proposed as part of Saudi Arabia's future energy mix. With recent interest in investing in small modular reactors, the Kingdom's strategy is both economically perplexing and risky. This paper shows that nuclear power, large and small, does not meet the criterion of economic competitiveness. First, nuclear electricity is already more expensive than that produced by solar technologies. The coupling of renewables and natural gas offers Saudi Arabia the most economically optimal option. Given that the Kingdom does not export its natural gas, the cross-over value between nuclear and natural gas was estimated at $13.6/mmBTU. For nuclear electricity to become a cheaper option, Saudi Arabia will have to sell its natural gas at higher prices than

372. A typical PV plant will need two persons/ MW in the construction stage, 7.5 person/ MW in the operation stage, and 0.5 person/ MW for maintenance. In 2012, the PV industry created 3-7 direct jobs and 10-20 indirect jobs per MW produced. *Energy from the Desert: Very Large scale PV Plants for Shifiting to Renewable Energy Value.* s.l. : IEA - PVPS T8-01, 2015.

that of the cross-over price. On the other hand, renewables, particularly solar power, offer Saudi Arabia real opportunities for meeting its electricity demand incrementally with substantial localization potential.

Saudi Arabia Energy Needs and Nuclear Power

Robin Mills

1. Introduction

This report, commissioned by the Nonproliferation Policy Education Center, assesses the prospects for Saudi Arabia to meet its energy and environmental requirements without the use of nuclear power.

There are evidently many ways in which those requirements could be defined and met. This study covers the period up to 2040 since this is the period over which the country has defined its current nuclear power target, and with a main focus on 2032 given that this was the year for its initial planned mix of nuclear and renewable energy. Reasonable assumptions are made for growth in electricity demand, other industrial energy demand, and transportation, based on official Saudi forecasts with independent checks and scenarios, including those with greater realized energy efficiency. The primary focus is on electricity given that this would be the main output of a civil nuclear power programme, so the other uses of oil and gas in the transportation and industrial sectors are not considered in detail.

Then a comparison is made of the Saudi energy system up to 2040 with nuclear power at current targeted levels; with the current fossil-fuelled mix; and with a number of scenarios with more renewable energy and no nuclear. These three approaches are contrasted on the metrics of economic attractiveness (cost and value); environmental outcomes; reliability and energy security; and local economic development. Possible proliferation-related reasons for pursuing civilian nuclear power have been dealt with at length by others[373] and are not considered here.

The main scenarios considered assume that the regional political situation and climate remain broadly similar to today. It is assumed that, over this period, transport (air, land and sea) will remain fuelled primarily by hydrocarbons (oil and possibly natural gas). In all scenarios, mostly current technologies are assumed, with conservative views on improvements in performance and cost. Of course, more rapid advances in renewable energy, unconventional hydrocarbons, carbon capture & storage, advanced energy storage or other technologies would make it easier to meet the country's energy requirements without nuclear power. Conversely, advances in nuclear power (such as small modular reactors) would make it relatively more attractive.

Environmental requirements are assumed to be, at minimum, compliance with Saudi Arabia's Paris climate change agreement submission, as well as mitigating local environmental impacts of its current energy system.

2. Current Situation

2.1. Objectives

Saudi Arabia's oil and gas are its main natural resource, the foundation of its economy and its political system and international

373. Johnson, K., 2019. *Who's Afraid of Saudi Nukes?.* [Online] Available at: https://foreignpolicy.com/2019/02/22/whos-afraid-of-saudi-nukes-123-kashoggi-mbs-russia-china/ [Accessed 23 May 2019]; Bowen, W. & Moran, M., 2015. Living with nuclear hedging: the implications of Iran's nuclear strategy. *International Affairs,* 91(4), pp. 687-707.

role. With that in mind, the key goals of Saudi energy policy can be defined as:[374]

1. Maintain and increase oil and gas output to meet the needs of the domestic population, supply local business with energy and feedstock, and earn export revenues
2. Maintain its role as one of the world's leading oil producers and the dominant force in the Organisation of Petroleum Exporting Countries (OPEC)
3. Secure markets for its oil by investing in assets and relationships in its main customers
4. Diversify the economy by developing hydrocarbon-based and energy-intensive industries, as well as, more recently, non-oil industries, with a view towards the long-term diminution of oil's role in the global economy
5. Build up substantial fiscal reserves to safeguard against oil price volatility
6. Safeguard the security of its energy industry and country generally via hard security measures, relationships/alliances and "soft power"
7. Delay or minimise environmental measures that would reduce the use of oil in the world economy

More recently, two other objectives have risen to prominence, even if they do not yet eclipse the objectives above:

1. Improve energy efficiency, introduce non-hydrocarbon energy sources, and reform subsidies to reduce the fiscal burden on government and preserve oil and gas for industrial and export use, while bearing in mind domestic public opinion and preserving the competitiveness of business

---

374. Author's interpretation based on observable Saudi policies, official statements and development plans such as Vision 2030

2. Improve environmental performance, including greenhouse gas emissions

2.2. Oil

Saudi Arabia is, of course, one of the world's largest oil producers (11.95 million barrels per day of crude oil, condensate and natural gas liquids in 2017, second in the world behind the USA and ahead of Russia; of this, 9.954 million bbl/day was crude oil), and the world's largest oil exporter (net 8.4 million barrels per day in 2017;[375] of this, 7.0 million bbl/day was crude oil and 1.4 million bbl/day refined petroleum products, with 0.49 million bbl/day import of refined products)[376]. It is the world's largest holder of conventional oil reserves with an official total of 266.2 billion barrels at the end of 2017[377] (only Venezuela has a larger figure, but the majority of Venezuela's official reserves are extra-heavy oil of uncertain commerciality).

Saudi oil production costs are exceptionally low, probably on the order of $10 per barrel or less (capital plus operating costs). The reserves/production ratio is 59 years, indicating (simplistically) that production could continue at current rates for 59 years from the existing reserves base. Saudi Arabia has aspirations to raise its total recoverable oil substantially, mostly by increasing recovery from the known fields.

Saudi Arabia is also a major oil consumer, refining 2.9 million bbl/day in 2017 and consuming 3.9 million bbl/day,[378] the fifth

---

375. BP Statistical Review of World Energy 2018, 'Oil – proved reserves', 'Oil Production – Barrels', 'Oil Consumption – Barrels'
376. OPEC Annual Statistical Bulletin 2018, p8
377. OPEC Annual Statistical Bulletin 2018, p8
378. OPEC Annual Statistical Bulletin 2018, p8

largest figure in the world (behind the US, China, India and Japan). Of this, a large proportion was crude oil burnt directly in power plants. Figure 1 shows oil consumption by month during 2013-19, and it can be seen that it rises some 700-800,000 barrels per day in summer versus winter, mostly due to an increase in the direct burning of crude oil and to some extent fuel oil. From 2016 onwards, direct use of crude, and consumption of diesel, has been cut by substitution with gas and some fuel oil.

Figure 1 Saudi Arabia monthly oil consumption[379]

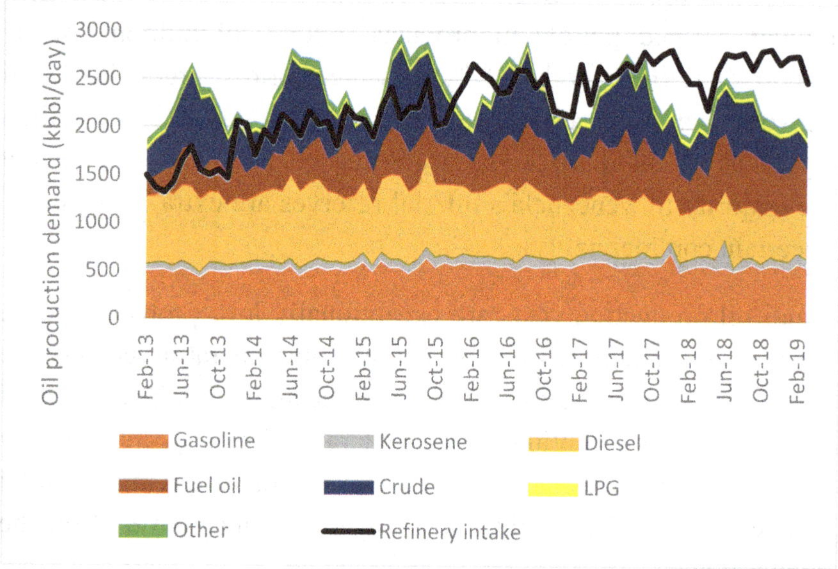

All Saudi oil and gas production, apart from that in the Neutral Zone with Kuwait, is operated by Saudi Aramco, the state-owned oil company. Aramco also operates the Kingdom's oil refineries (some in partnership with foreign companies), has a substantial

379. Qamar Energy graph from Joint Organisations Data Initiative (JODI)

petrochemical and power generation business, and has a number of significant international investments in refining and petrochemicals in the US, China, South Korea, Japan, and elsewhere. Aramco is respected as a relatively efficient and technically proficient operator. There are plans for an Initial Public Offering (IPO) of 5% of Aramco on the Saudi stock exchange (the Tadawul) and a to-be-chosen international exchange, which have been repeatedly delayed and now are apparently set for 2021. Aramco recently acquired a 70% stake in Saudi Arabia Basic Industries Corporation (SABIC), a large petrochemical producer, from the Public Investment Fund, the Saudi sovereign wealth entity.

2.3. Natural Gas

Saudi Arabia is also an important producer and consumer of natural gas, though not to the same extent as for oil. Official reserves at the end of 2017 were 283.8 trillion cubic feet (the 6th-largest in the world), and marketed production was 10.8 billion cubic feet per day (the 8th largest in the world). All of this gas was used domestically, making the Saudi domestic gas market the world's 6th largest.

A large part of gas production is associated (a by-product of oil production) and is therefore cheap to collect and use but not flexible. The proportion of associated gas, though, has fallen from 42% in 2005 to about 33% in 2015.[380] The reserves/production ratio for gas is 77 years, indicating ample reserves to increase production; however, if oil production remains constant, significant increases in gas production would have to come from non-associated fields. The country has developed its first non-associated fields offshore in the Gulf, Karan and (together feeding the Wasit gas plant) the Arabiyah

---

380. Jadwa Investment, 2016. Natural Gas and the Vision 2030. *Jadwa Investment Economic Research*, October, p. 3.

and Hasbah fields, and the small onshore Midyan field in the north-west.

Saudi Arabia has significant unconventional (tight/shale) gas resources, estimated at 645 trillion cubic feet.[381] Apart from initial projects in the north-west of the country (Tabuk Basin), shale gas production has consisted only of pilots, with estimates of high production costs ($9/MMBtu in 2012). However unconventional gas production is targeted to reach 4 Bcf/day by 2026,[382] with a particular focus on the Jafurah Basin in eastern Saudi Arabia, which has been compared to the Eagle Ford formation of South Texas. Most versions of hydraulic fracturing, required to produce tight gas, use fresh water, which would require desalination. BP has developed tight reservoirs in Oman's Khazzan field, also in a desert setting, using reverse osmosis desalination plants. Alternatively fracturing fluids using saline water, propane or carbon dioxide could be employed though they may be less effective. In either case, this raises the likely initial cost of tight gas production in Saudi Arabia.

Although it borders two of the world's largest gas resource holders, Iran and Qatar, Saudi Arabia by policy does not import or export gas. Difficult political relations with these two countries (and with another gas-rich neighbour, Iraq, although relations here have somewhat improved recently) make direct imports unlikely. Other than Yemen, whose moderate LNG exports are shut down by the continuing war, its other neighbours – Egypt, Jordan, UAE, Oman, Bahrain, and Kuwait – are themselves gas importers or about to become so.[383] If political conditions did change to allow gas imports

---

381. Baker Hughes, quoted in
Oil & Gas News, 2019. Work starts on Shale Production. *Saudi Arabia Review,* June.36(6).
382. Hart Energy, 2016. *Saudi Shale Drive Powers Up.* [Online]
Available at: https://www.hartenergy.com/exclusives/saudi-shale-drive-powers-28752#p%3Dfull [Accessed 29 June 2019].
383. Oman exports LNG but the plant has run below capacity because of a shortage of feedstock; it has negotiated with Iran for gas imports. Increased

from a neighbour, its lower cost might undermine high-cost domestic gas production. It is also possible that Saudi Arabia could at some point import liquefied natural gas (LNG).[384] In July 2017, Russian energy minister Alexander Novak said that Aramco had been offered a share in Novatek's Arctic LNG-2 project in the north of west Siberia.[385]

Energy minister Khaled Al Falih has indicated an aspiration to raise gas from 50% of the "energy mix" to 70%, which probably refers to its share of electricity generation (gas's share of total primary energy use is rather lower, at about 32% in 2016). Under the National Transformation Plan, gas production is intended to rise from a (gross) 12 Bcf/day in 2016 to 17.8 Bcf/day in 2020 and 23 Bcf/day by 2026.[386] Gas production is shown in Figure 2, as is the projected production in 2020 and 2026, assuming the proportion of shrinkage (removal of natural gas liquids) remains the same (shrinkage figures are not available for 2016). On official figures, flaring and reinjection is minimal although likely to be substantially more in reality. It can be seen that, although marketed gas production has increased by 2.3 Bcf/day from 2011 to 2017, it has to increase by a further 7 Bcf/day by 2020. 2.5 Bcf/day of this will come from the Fadhili processing plant due online in 2019 (which will process 0.5 Bcf/day from the onshore Khursaniyah field and 2 Bcf/day from the second phase of the offshore Hasbah field), and 1.3 Bcf/day from an

---

unconventional gas production has now allowed it to run the plant at full capacity again. Egypt has again become a gas exporter with development of its Zohr field, but any excess is likely to continue to be exported through previously mothballed LNG plants on the Mediterranean coast.

384. Reuters, 2016. *Saudi Aramco could import gas to boost use in energy mix: minister.* [Online] Available at: https://www.reuters.com/article/us-saudi-plan-saudi-aramco-gas-int-idUSKCN0YS2MS [Accessed 07 June 2016].

385. Reuters, 2017. *Russia offers Saudi Aramco role in Arctic LNG project: Novak.* [Online] Available at: https://www.reuters.com/article/us-russia-saudi-novak-idUSKBN19R1RF [Accessed 06 July 2017].

386. Hart Energy, 2016. *Saudi Shale Drive Powers Up.* [Online] Available at: https://www.hartenergy.com/exclusives/saudi-shale-drive-powers-28752#p%3Dfull [Accessed 29 June 2019].

expansion of the Hawiyah gas plant.[387] 0.075 Bcf/day will be produced from the Midyan field, in a new gas-producing area in the north-west of the country. It is not clear where the remainder of the increase, 1.125 Bcf/day, will come from, although an expansion of the Haradh gas processing plant is also in the works (both Hawiyah and Haradh process gas from the Ghawar field).

Figure 2 Gas production[388]

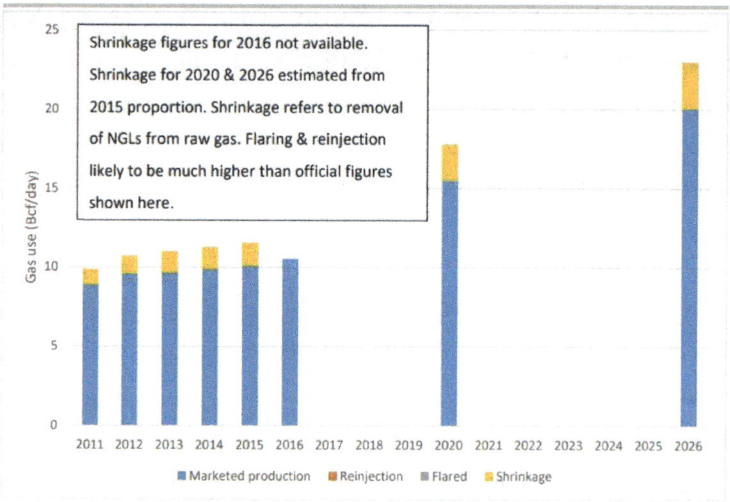

The International Energy Agency has become progressively more skeptical on Saudi Arabia's gas output targets. In 2014, it forecast production would be approaching 13 Bcf/day by 2019, but in subsequent editions of its Medium Term Gas Outlook, it reduced that down to just over 9 Bcf/day by 2019 and about 10 Bcf/day by

387. Reuters, 2017. *Saudi Aramco to boost gas production at Hawiyah, Haradh: sources.* [Online] Available at: https://www.reuters.com/article/us-aramco-gas-idUSKBN1541PK [Accessed 20 January 2017].
388. OPEC, 2016. *OPEC Annual Statistical Bulletin,* Vienna: Organization of the Petroleum Exporting Countries.; BP Statistical Review of World Energy 2018 for year 2016

2022.[389] This has proved to be pessimistic, given BP's figures for 2017 production of 10.8 Bcf/day (though the IEA has, for some reason, a lower baseline).

Gas consumption by sector is shown in Figure 3. The power sector is about 56% of total gas consumption. (Note that there is an unexplained discrepancy between the IEA figures, used here, and the OPEC and BP figures for total production, which are significantly higher, while Saudi Electricity Company (SEC)'s figures for generation use are significantly lower than those of the IEA). 26% of the country's petrochemical feedstock is methane (the main constituent of natural gas) and 46% is ethane (primarily derived from natural gas).[390]

Figure 3 Gas consumption by sector[391]

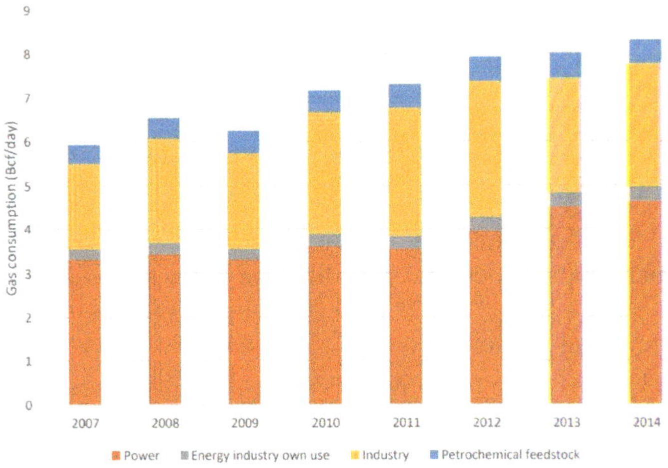

389. International Energy Agency, Medium Term Gas Report 2014, 2015, 2016, 2017; Middle East Economic Survey volume 60 number 28 (14th July 2017), p2
390. Jadwa Investment, 2016. Natural Gas and the Vision 2030. *Jadwa Investment Economic Research*, October, p. 1-15.
391. IEA, 2015. *Natural Gas Production.* [Online] Available at: https://www.iea.org/statistics/?country=SAUDIARABI&year=2015&category=

Gas prices are regulated by the government and are very low by global standards, being raised in 2016 from $0.75 to $1.25/MMBtu, and for ethane (used in petrochemicals) from $0.75 to $1.75/MMBtu.[392] These are still well below the price that would be paid for imported LNG, or the cost of developing domestic unconventional gas. In Saudi Aramco's bond prospectus of 2019, the prices paid to Aramco the Saudi government by from 2020 onwards are listed as $0.31/MMBtu for associated gas, $1.52/MMBtu for southern non-associated gas, $3.84/MMBtu for northern non-associated gas, and $3.8/MMBtu for Fadhili non-associated gas . However these prices may not be the same as those consumers are then charged for that gas.

2.4. Electricity

Electricity demand has been rising rapidly, at an annual average rate of 6.3% in the ten years 2007-17, spurred by the economic boom (up to 2014), rising population, and low, subsidised tariffs. However, with an economic slowdown and an increase in tariffs, demand grew by only 1.4% in 2017. It remains to be seen whether this slower growth becomes the norm, but without efficiency measures the government expects demand growth to be 4-5% annually.[393]

Total generating capacity attached to the national grid was 83.1 GW at the end of 2018 (up from 69.2 GW in 2015), of which 58 GW was provided by the state-owned Saudi Electricity Company and the rest

---

Natural%20gas&indicator=NatGasProd&mode=chart&dataTable=GAS [Accessed 22 October 2016].

392. Oxford Business Group, 2016. *Saudi Arabia plans to diversify power sources.* [Online] Available at: https://oxfordbusinessgroup.com/news/saudi-arabia-plans-diversify-power-sources [Accessed 09 September 2016].

393. Oxford Business Group, 2016. *Saudi Arabia plans to diversify power sources.* [Online] Available at: https://oxfordbusinessgroup.com/news/saudi-arabia-plans-diversify-power-sources [Accessed 09 September 2016].

by industrial companies with their own grid-connected power plants. The highest reported load was 62.26 GW in August 2015, and peak load was lower in 2016-18, although that may be exceeded this year.[394] The reserve margin is therefore 33%, which would be considered more than adequate (SEC's target is 12%). Power plants receive fuel at prices fixed by the government, which are far below world market levels: $1.25 per MMBtu for natural gas, $4.4 per barrel for heavy fuel oil, $11.7 per barrel for diesel, and $5.94 per barrel for crude oil (these are the prices after a sharp increase ordered in January 2016; they were even lower before).[395]

Figure 4 shows Saudi power generation by source. It can be seen that virtually all comes from oil (crude oil and refined products) and natural gas. A very small proportion (0.135 TWh out of total generation of 375.6 TWh in 2017, or 0.036%) is provided by solar photovoltaics (PV), with current installed capacity of 85 MW.[396] Saudi Arabia's first wind turbine (2.75 MW) was installed in January 2017[397] and it does not currently make use of any other form of renewable electricity production,[398] nuclear power, or coal. Saudi Arabia is constructing three integrated solar combined cycle (ISCC) plants, using small amounts of solar thermal to boost the efficiency of combined-cycle plants: Duba 1 (43 MW solar out of total 605 MW), Waad Al Shamal (50 MW solar out of total 1390 MW), and

---

394. Saudi Electricity Company

395. Middle East Economic Survey (8th January 2016), Volume 59, Number 1, p8; Walid Matar and Murad Anwer (October 2017), 'Jointly reforming the prices of industrial fuels and residential electricity in Saudi Arabia', Energy Policy, Volume 109, p747-756

396. MESIA, 2016. *Middle East Solar Outlook for 2016,* Dubai: MESIA.

397. Saudi Aramco, 2016. *Saudi Aramco and GE deliver the first wind turbine in the Kingdom of Saudi Arabia.* [Online] Available at: https://www.saudiaramco.com/en/news-media/news/2016/ge-turbine [Accessed 18 December 2016].

398. There are some solar hot water systems.

Taiba (180 MW solar out of total 3600 MW).[399] The 399 MW Sakaka solar PV project is under construction and intended to start generation in late 2019.

KA CARE's 2013 strategy foresaw total generation capacity by 2032 of 123 GW. This would be met by 60 GW of hydrocarbon power (oil and gas), 18 GW of nuclear, 41 GW of solar (16 GW PV and 25 GW CSP), 3 GW of waste-to-energy, and 1 GW of geothermal. A further 9 GW of wind power would be used for desalination, with the intermittency presumably managed by other generation and/or water storage.

80% of the renewable and 60% of the nuclear expenditure would be in-country (local content). By 2020, 17.35 GW of solar and 6.5 GW of other renewables would have been installed.

However, very little progress was made on this plan, due to bureaucratic infighting and unclear lines of responsibility.[400] In 2016, the new National Transformation Plan intended that 3.45 GW of renewable energy should be introduced by 2020 (4% of the energy mix) and that the renewable share would reach 9.5 GW[401] by 2023 and 10% by 2030 (instead of almost 50% in the KA CARE plan). In April 2017, companies were shortlisted for the 300 MW Sakaka solar plant, with award made to private Saudi developer

---

399. Hashem, H., 2016. *Saudi Arabia's first ISCC plant to set record low CSP cost.* [Online] Available at: http://tracker.newenergyupdate.com/technology/saudi-arabia%E2%80%99s-first-iscc-plant-set-record-low-csp-cost [Accessed 11 November 2016].
400. Apricum, 2016. *Potentially Game-Changing Saudi Arabian Government Restructuring Bolsters 9.5 GW Renewable Energy Target by 2023.* [Online] Available at: https://www.apricum-group.com/saudi-arabia-announces-9-5-gw-renewable-energy-target-new-king-salman-renewable-energy-initiative/ [Accessed 9 May 2016].
401. Tsagas, I., 2017. *WFES 2017: Saudi Arabia announces new energy master plan.* [Online] Available at: https://www.pv-magazine.com/2017/01/17/wfes-2017-saudi-arabia-announces-new-energy-master-plan/ [Accessed 17 January 2017].

Acwa Power in February 2018, and the 400 MW Dumat Al Jandal wind farm. The wind power, originally intended to be at Midyan in the north-west Tabuk region, will now be in Al Jouf in the north,[402] and by July 2017 was already five months behind its original target. In April 2019, bids were received for Dumat Al Jandal, setting world record low levels for onshore wind.

Under the latest plans from the Renewable Project Development Office (REPDO) of the Ministry of Energy, released in February 2019, 3.4 GW of renewables will be tendered in 2019 (1.515 GW of solar PV tendered in January, with a further 1010 MW of solar PV and 850 MW of wind to be tendered later in the year). Renewable capacity would reach 27.3 GW by 2023 and 59-60 GW by 2030, both highly ambitious targets given the pace of progress so far. The 59 GW target would be made up of 40 GW of solar PV, 16 GW of wind and 3 GW of solar thermal (CSP). At the same time, the share of gas in generation would rise to 70% (with 30% renewables), so oil would be completely eliminated from the generation mix. This is challenging given that some new oil-fired power plants are still under construction and that western Saudi Arabia currently has no access to gas supplies except at Yanbu' and in the far north-west for mining projects.

---

402. Graves, L., 2017. *Saudi Arabia looks past oil and pushes ahead with wind energy ambitions.* [Online] Available at: https://www.thenational.ae/business/energy/saudi-arabia-looks-past-oil-and-pushes-ahead-with-wind-energy-ambitions-1.609330 [Accessed 16 July 2017].

Figure 4 Saudi Arabia electricity generation by source[403]

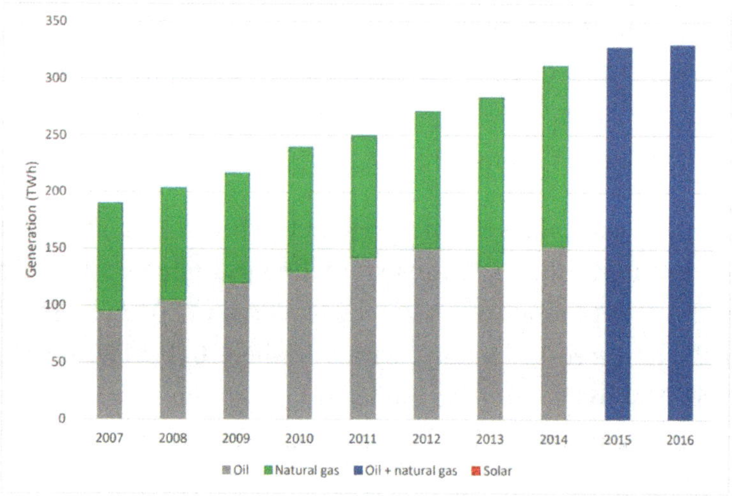

Figure 5 Fuel to electricity generation[404]

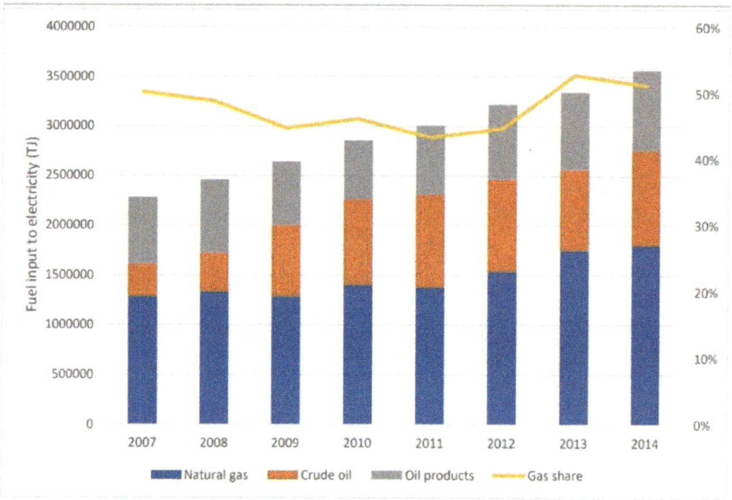

403. BP Statistical Review of World Energy 2018, 'Solar Consumption – TWh', 'Electricity Generation'; International Energy Agency. *Statistics.* [Online] Available at: https://www.iea.org/statistics

404. Data from International Energy Agency. *Statistics.* [Online] Available at: https://www.iea.org/statistics

Figure 5 shows a breakdown of fuel supply to power generation. 2322614 TJ of gas supplied in 2016 is equivalent to about 5.97 Bcf per day. The share of gas in the fuel mix has hovered a little below 50%, rising to 62% by 2016 as new gas supplies became available. Saudi Arabia is almost unique globally for the large-scale direct use of crude oil in power generation, which averaged about 450 kbbl/day during 2014 and reached 1 million bbl/day at times during the summer months when demand for air-conditioning peaks.

Figure 6 Power generation efficiency by fuel[405]

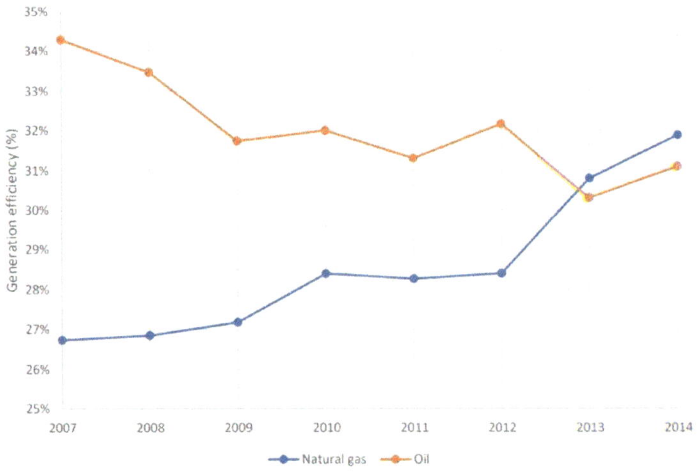

Figure 6 shows gross power generation efficiency for oil (crude and products) and natural gas. The efficiency of gas-fired power plants has risen somewhat over time, while that of oil-fired power plants has declined. But the overall efficiency of both is low. SEC claims that the efficiency of its plants was 35.6% in 1Q 2017, compared to

405. Calculated from data from International Energy Agency. *Statistics.* [Online] Available at: https://www.iea.org/statistics

34.5% in 1Q 2016[406], although as this period is cooler it would be expected that efficiency would be higher than later in the year. Nevertheless as this figure is higher than the overall efficiencies shown in Figure 6, this implies that the non-SEC power generation has low efficiency.

The hot and (along the Red Sea and Gulf coasts) humid climate, the highly seasonal load profile (peak demand in summer), and, for cogeneration plants, the requirement to co-produce desalinated water all reduce efficiency from its theoretical maximum. Nevertheless, these efficiency factors are still very low, as over 50% should be achievable in realistic Gulf operating conditions for a modern combined-cycle plant running on natural gas. Oman, for example, has raised the overall efficiency of its generating fleet to 40% in recent years.

Demand is highly seasonal due to air-conditioning loads (see Figure 7), with a peak daytime summer load of about 60 GW and a summer night-time load of about 50 GW, contrasting to about 35 GW daytime and 25 GW night-time in winter.

---

406. Saudi Electricity Company

Figure 7 Annual demand pattern, day and night[407]

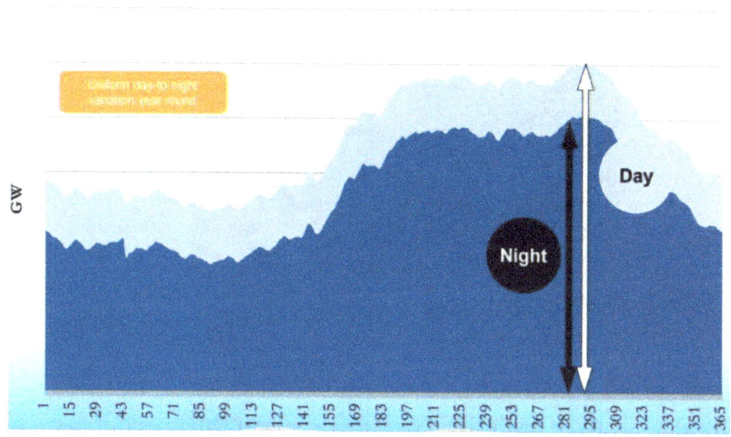

The daily demand pattern is shown in Figure 8. The peak in summer is reached in mid-afternoon, with maximum temperatures and hence air-conditioning load, exacerbated by poor insulation, inappropriate building techniques (e.g. glass-paneled office buildings) and use of inefficient air-conditioning units. There is a second peak around 20-21:00 when temperatures remain high and demand is driven by residential use as people return from work. Even at night consumption does not drop much, as temperatures remain high, falling only by around 6 am.

---

407. Ghabban, A. A., 2013. *Saudi Arabia's Renewable Energy Strategy and Solar Energy Deployment Roadmap,* Riyadh: KACARE.

Figure 8 Forecast daily load pattern, 2032[408]

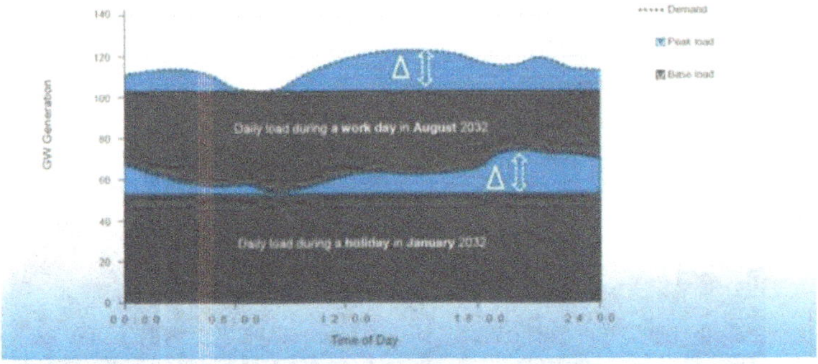

As noted, Saudi electricity prices to consumers are highly subsidised. In 2015, the average countrywide price was 2.09 US₡/kWh, with tariffs tiered by consumption, from the lowest band of 2000 kWh up to amounts in excess of 10 000 kWh. Price rises in 2016 increased the price in the main band by 20%, to 4.3 US₡/kWh.[409] However, this remains well below international benchmarks and the cost of generation (if fuel were valued at international prices). These low prices encourage wasteful consumption and make energy efficiency investments or small-scale renewable energy (such as rooftop solar panels) unattractive. SEC is vertically-integrated, and so it is not clear what proportion of its costs relate to transmission and distribution (T&D), or how the end-user subsidy is distributed between generation and T&D. Tariffs do not vary by location, and therefore remote areas are effectively being undercharged for T&D. If all costs were transparently accounted for, local renewable energy may be more cost-competitive for such localities.

---

408. King Abdullah Centre for Atomic and Renewable Energy (KA CARE)
409. Jadwa Investment, 2016. Natural Gas and the Vision 2030. *Jadwa Investment Economic Research*, October, p. 1-15.

Saudi Arabia has an extensive electricity grid that covers essentially the entire country apart from some uninhabited parts of the Rub' Al Khali (see Figure 9). Some remote or rural areas are served by unconnected local generation (usually diesel generators) but all of the population has access to electricity. As can be seen from the map, the east-west connections are limited with only two in operation, but more are under construction or planned.

Figure 9 Saudi electricity grid[410]

Saudi Arabia is connected to the five other Gulf Cooperation Council (GCC) states via the GCC Interconnection for electricity. This has a capacity of 1200 MW to Kuwait, 1200 MW from Saudi Arabia into the interconnection, 600 MW to Bahrain, 1200 MW to Qatar and 1400 MW to the UAE (with 400 MW to Oman through

410. Saudi Electricity Company, 2016. *National Grid Marketing Plan,* Riyadh: SEC.

the UAE).[411] Construction of a 3000 MW interconnection with Egypt is also underway with first operations scheduled for the first quarter of 2020 , and a connection to Jordan is under discussion.

3. *Nuclear Power in Saudi Arabia*

Saudi Arabia's most recent interest in nuclear power dates back to December 2006, when the GCC states (Saudi Arabia, Bahrain, Kuwait, Oman, Qatar, and the United Arab Emirates) announced the commissioning of a study on civil nuclear power. A number of different plans have emerged since then, as described below.[412] This list is not exhaustive, and other agreements have also been concluded on research, training, etc. However, Saudi Arabia has not concluded a "123 agreement," which is required for U.S. companies to export nuclear equipment or materials to it.[413]

August 2009 – Government announcement of a national nuclear power programme

2009 – Safeguards agreement in place with the IAEA, but no Additional Protocol

April 2010 – Royal Decree announces the establishment of the King Abdullah Centre for Atomic and Renewable Energy (KA-CARE) to supervise nuclear and renewable programmes.

---

411. Al-Shahrani, N., 2015. *GCC Interconnection Super Grid; Launch Pad of GCC Energy Power Market.* [Online] Available at: http://www.epcc-workshop.net/Presentations/EPCC2015-Presentation13_Al-Shahrani-GCCIA.pdf [Accessed 2016].
412. Material in this list is based primarily on http://www.world-nuclear.org/information-library/country- profiles/countries-o-s/saudi-arabia.aspx, with other sources noted where used.
413. Nuclear Energy Institute, 2016. *Compete Globally.* [Online] Available at: https://www.nei.org/advocacy/compete-globally

2011-13 – Various plans and targets announced, including site selection for reactors at Jubail on the Persian Gulf and Jizan and Tabuk on the Red Sea.

June 2011 – Plans for 16 reactors over the following 20 years (likely of ~1 GW each), costing $80 billion, generating 20% of Saudi Arabia's electricity, with smaller reactors used for desalination.

April 2013 – KA-CARE proposes 17 GW of nuclear power by 2030, plus renewable energy (see separate section). Construction was to begin in 2016.

September 2013 – GE Hitachi Nuclear Energy signed a contract with Exelon Nuclear Partners to pursue construction of its Advanced Boiling-Water Reactor (ABWR) and (ESBWR), while Toshiba/Westinghouse signed a deal with Exelon to pursue its AP1000 and ABWR reactors with KA- CARE.

January 2015 – Nuclear target date moved to 2040.

March 2015 – Korea Atomic Energy Research Institute (KAERI) signs an agreement with KA-CARE to assess building two or more SMART reactors (up to 100 MW each), including integrated desalination, with a cost for the first unit of up to $1 billion. Construction was to begin in 2018.

April 2015 – Government adviser Maher Al Odan says the country has a target of 6-7 GW of nuclear power by 2032 and 17 GW by 2040.[414]

November 2015 – Announcement at a conference that Saudi Arabia has identified a number of possibly commercial uranium deposits in

414. Rascouet, A. & Mahdi, W., 2016. *Saudi Arabia to Select Nuclear Power-Plant Site 'Very Soon'.* [Online] Available at: https://www.bloomberg.com/news/articles/2016-10-20/saudi-arabia-to-select-nuclear-power-plant-site-very-soon [Accessed 20 October 2016].

its territory. These include volcanic, black shale, evaporite and possibly pegmatite deposits.[415]

January 2016 – KA-CARE signs with China Nuclear Engineering Corporation (CNEC) for a high- temperature reactor to be built in Saudi Arabia, based on the HTR-PM being built in China by CNEC.

September 2016 – Russia's Rosatom announces that it is ready to build 16 nuclear reactors in Saudi Arabia at a cost of $100 billion.[416]

October 2016 – Saudi energy minister Khalid Al Falih announces that sites for the first nuclear plant will be chosen within the next 12 months.

January 2017 – Al Falih says that the first two nuclear reactors, totalling 2.8 GW, were in the early stages of feasibility and design studies.[417]

March 2017 – Further agreement signed with CNEC on localisation of the nuclear supply chain in Saudi Arabia. Saudi Arabia announces an investment fund for its neighbour and ally Jordan, which includes uranium mining.[418]

---

415. Howari, F. M., Goodell, P. & Abdulaty, S., 2009. *Uranium Resources in the Middle East.* [Online] Available at: https://www-pub.iaea.org/mtcd/meetings/PDFplus/2009/cn175/URAM2009/Session%203/05_88_Howari_USA.pdf
416. Arabian Business, 2016. *Russia's Rosatom says ready to build 16 Saudi nuclear power stations.* [Online] Available at: https://www.arabianbusiness.com/russia-s-rosatom-says-ready-build-16-saudi-nuclear-power-stations-644543.html
417. Reuters, 2017. *Saudi Arabia Says It's About to Launch a $30-$50 Billion Renewable Energy Program.* [Online] Available at: https://fortune.com/2017/01/16/saudi-arabia-renewable-energy-program/
418. AFP, 2017. *Saudi Arabia to set up $3bn joint investment fund for Jordan.* [Online] Available at: https://www.thenational.ae/world/saudi-arabia-to-set-up-3bn-joint-investment-fund-for-jordan-1.48544

May 2017 – Program framework agreement signed with IAEA; KA CARE holds discussions with China.

24th July 2017 – National Panel for Atomic Energy established.[419] KA CARE said to be undertaking a technical study for the location of two large and two small reactors, and to explore for uranium for mining and processing its own fuel.[420]

January 2018 – Two sites on the Gulf coast, Umm Huwayd and Khor Duweihin, shortlisted for the first two reactors.

July 2018 – IAEA team visited Saudi Arabia to review its nuclear power programme .

December 2018 – Rosatom organized a workshop in Riyadh on Russian nuclear power technologies .

March 2019 – KA CARE set up a joint programme with the Jordan Atomic Energy Commission and the Jordan Uranium Mining Company.

April 2019 – Research reactor (30 kW) near Riyadh making rapid progress on construction.

As can be seen from this timeline, Saudi Arabia's nuclear plans so far have changed frequently and involve numerous outside parties without clear progress, and with unrealistically fast timelines. However, this is not unique to its nuclear sector, as the renewable energy programme championed by KA-CARE has made similarly limited progress. Some recent activity may indicate that the pace of both the nuclear and renewable schemes is finally picking up, but it remains to be seen how effectively the new 'super- ministry' of

419. Arab News, 2017. *Yamani lauds Cabinet approval for establishment of National Atomic Energy Project.* [Online] Available at: http://www.arabnews.com/node/1134776/saudi-arabia
420. Middle East Economic Survey Volume 60, Number 30, p.7

energy and industry, under Khalid Al Falih, will be able to coordinate these complex activities while managing other initiatives such as the now-delayed IPO of part of Saudi Aramco, the acquisition of Saudi Basic Industries Corporation (SABIC) by Aramco, the development of the mining sector, and the privatisation of Saudi Electricity Company.

4. Environment

Saudi Arabia's Intended Nationally Determined Contribution (INDC)[421] under the 2015 Paris climate agreement seeks to mitigate up to 130 million tonnes of CO2 equivalent by 2030.[422] This is to be achieved by economic diversification, energy efficiency (including the conversion of single-cycle power plants to combined cycle), the use of renewable energy, carbon capture and storage, increased use of gas (by implication, to replace oil), reducing flaring, water management, public transport, reduced desertification and other measures.

The submission does not give a figure for estimated business-as-usual emissions by 2030, but estimated CO2 emissions (excluding other greenhouse gases) from fuel combustion alone were 570.4 million tonnes in 2014, 590.8 million tonnes in 2016[423] and 594.7 million tonnes in 2017. The proposed mitigation is therefore about 22% of 2017 emissions. For electricity generation only, CO2

421. UNFCCC, 2016. *INDCs as communicated by Parties.* [Online] Available at: https://www4.unfccc.int/sites/submissions/indc/Submission%20Pages/submissions.aspx
422. Including carbon dioxide, and other gases such as methane and nitrous oxides at their CO2-equivalent warming potential.
423. BP Statistical Review of World Energy 2018, 'Carbon Dioxide Emissions'

emissions in 2016[424] were approximately 130 million tonnes from natural gas, 68 million tonnes from crude oil and 60 million tonnes from oil products, totalling 258 million tonnes.[425] In other words CO2 emissions from electricity generation were about 44% of the total in 2016, having risen from 40% in 2014.

5. Energy Outlook

5.1. Forecasts

5.1.1. Demand

Total generation was 375.6 TWh in 2017[426] and peak load was 62.12 GW (including about 2.3 GW from remote, non-grid-connected locations)[427]. As noted, peak load was assumed by KA CARE to grow to 123 GW in 2032 plus another 9 GW required for desalination. Including the desalination load, that is an implied growth rate of 5% annually. The growth rate over the past ten years has been 7.4% annually.[428] However a slower rate of growth in future is reasonable given increases in electricity tariffs, efforts to promote efficiency, and the slowing of economic growth due to lower oil prices.

---

424. Note there is a considerable unexplained discrepancy between EIA and Saudi Electricity Company figures
425. Author's calculation from IEA consumption figures and emissions factors from IPCC. *EFDB: Emission Factor Database.* [Online] Available at: https://www.ipcc-nggip.iges.or.jp/EFDB/main.php
426. BP Statistical Review of World Energy 2018, 'Electricity Generation'
427. ECRA. *Station Capacities.* [Online] Available at: https://www.ecra.gov.sa/en-us/DataAndStatistics/NationalRecord/StationCapacities/Pages/Home.aspx
428. BP Statistical Review of World Energy 2018, 'Electricity Generation'

If the pattern of demand has not changed by 2032, then that implies total required generation would be 716.5 TWh, peak daytime summer load would be 132 GW, night-time summer load would be about 110 GW, daytime winter load would be about 77.8 GW, and night-time winter load about 56.1 GW. This ignores the possibility of switching the desalination load between different periods.

5.1.2. Generation Mix

Estimated generation costs for various generation technologies under Saudi conditions are shown below (Figure 10). These do not include transmission and distribution costs nor the cost of backup for intermittent resources. The fuel price assumed is $7 per MMBtu and the thermal efficiency assumed is 50% for combined-cycle gas turbines (reasonable for current best-in-class under Gulf conditions), 31% for steam turbines and 25% for gas turbines. For the purpose of calculating fuel requirements and CO2 emissions, thermal plants are assumed to burn exclusively natural gas.

The prices for solar CSP (with 8 hours of storage) and solar PV are consistent with bids received in the UAE for recent projects, with Dubai's first CSP project attracting a bid of 9.45 ₡/kWh for 200 MW, lowered to 7.3 ₡/kWh for an expanded 700 MW plant with 11-15 hours of storage. However, since the bid cost of solar CSP at 7.3-9.45 ₡/kWh is well below previous projects (such as Morocco's Noor 3 at 15.67 ₡/kWh, despite excellent site conditions),[429] it should be regarded with caution – the project in Dubai has not yet been completed. Assumed capital cost for CSP is $3.6 million per MW, while the reported cost for the Duba and Waad Al Shamal

429. Graves, L., 2017. *24-hour solar power is possible for Mena.* [Online] Available at: https://www.thenational.ae/business/24-hour-solar-power-is-possible-for-mena-1.40805

ISCCs is $1.6 million per MW, albeit providing only a small amount of the plant's total capacity. The solar PV prices are more reliable given that there have been several bids around that level, as well as other sunny countries in Africa and Latin America. It is likely that solar PV costs will continue to fall, increasing its advantage over all the other technologies. Saudi Arabia's Dumat Al Jandal 400 MW wind project was awarded at 2.13 ₡/kWh, significantly below the cost shown here, which may therefore be pessimistic if further projects are awarded at good sites.

There is little information on Saudi geothermal resources and costs were therefore taken for a typical U.S. project; the planned capacity is small anyway so makes little difference to the analysis.

The cost for nuclear, at about 9.1 ₡/kWh, is consistent with reported costs for the UAE's nuclear programme, although given likely cost escalation and some delays, it is probably an underestimate. It is also questionable whether Saudi Arabia would be able to plan and execute a nuclear programme as quickly and cost-effectively as the UAE, particularly given its ambitions for local content. The reported capital costs are well below those for U.S. reactors, and may represent a combination of low (or zero) land costs, lower permitting and legal costs, the absence of public opposition, lower costs for land and possibly some materials, low (government-backed) financing costs, aggressive bids from suppliers, and over-optimism.

Advanced nuclear designs may achieve costs significantly below this,[430] with a (wide) range from 3.6-9 ₡/kWh and an average of 6 ₡/kWh. Some designs for small modular reactors, being investigated by Saudi Arabia in cooperation with KAERI as noted above, are included in the study quoted, and also have potential to

430. Energy Reform Innovation Project, 2018. *What will Advanced Nuclear Power Plants Cost?*. [Online] Available at: https://www.innovationreform.org/wp-content/uploads/2018/01/Advanced-Nuclear-Reactors-Cost-Study.pdf

provide nuclear power at more competitive costs. Cogeneration of desalinated water using waste heat could also improve the overall economics. However, these plants are still in design and commercial demonstration plants are at least a decade away. The cost estimates are also for nth-of-a-kind (NOAK) plants, i.e. incorporating substantial learning from several prior constructions, and the first-of-a- kind (FOAK) plants would cost more.

The costs do not include any future cost reduction from research and development, and therefore the costs of solar and wind in particular can be anticipated to fall; as noted, nuclear costs might fall if advanced designs are developed. Saudi Arabia's initial large PV tender shortlisted two bidders at 2.342 ₡/kWh and 2.66 ₡/kWh, rejecting the lowest bid of 1.79 ₡/kWh for having insufficient local content. This shows that despite the relatively smaller size (300 MW) of its plant and the fact it was its first tender, it can receive bids similar to, even slightly lower than, those achieved in the UAE. These prices are anyway amongst the very lowest offered anywhere in the world. If Saudi Arabia seeks to achieve high local content targets, which was a major component of the original KA CARE plan, this would raise costs, although achieving wider development goals. On the other hand, its CSP conditions in the north-west are superior to the UAE's.

Figure 10 Generation costs under Saudi conditions[431]

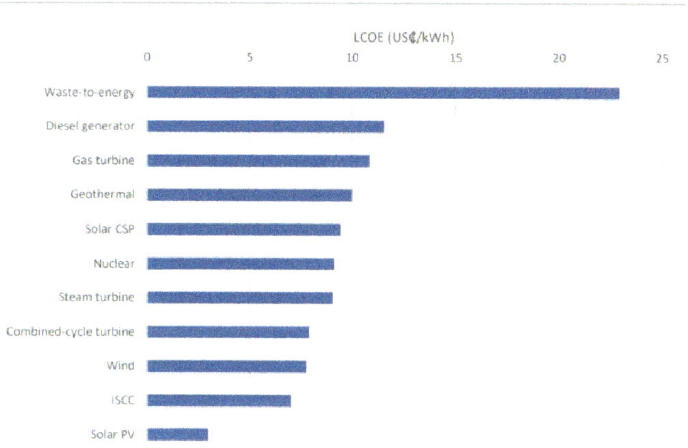

These costs cannot be viewed simplistically in isolation, because a realistic generation mix, as discussed below, requires a blend of generation methods. However, it does indicate that, at the relatively high fuel prices assumed (comparable to current LNG prices, and well above current regulated Saudi domestic prices), high-efficiency combined cycle gas turbines are still cheaper than nuclear or solar CSP, which have very similar costs in this view. If the lower CSP prices seen in Dubai could be replicated, or even reduced given the favourable resources in Saudi Arabia, much more CSP could be included given its capability for night-time storage. The CSP portion of ISCC plants, serving to boost their efficiency, is cheaper than a conventional CCGT and could be considered for inclusion in all suitably-located new CCGTs. Solar PV is extremely cheap, even if considered only as a means of saving fuel. Wind is also cheap and has the advantage of a different daily and seasonal pattern from solar. Some portion of wind could be taken as firm generation (i.e.

431. Qamar Energy research; EIA

not requiring routine backup), depending on the consistency of wind speeds.

60% of household water consumption comes from desalination, with the remainder from groundwater, and desalination demand is growing at 14% per year.[432] This study has not assessed the provision of desalination in detail. One study[433] has concluded that reverse osmosis plants driven by solar PV or CSP would be slightly cheaper than reverse osmosis driven by nuclear power, and that in any case all the reverse osmosis methods are much cheaper than the multi-effect distillation or multi-stage flash which currently predominates (and which is energy-intensive). It can also be observed that reverse osmosis is more flexible as it can be run when surplus electricity is available and hence function as a kind of energy storage/demand-shifting. Thermal desalination methods can be run using the waste heat from fossil fuelled or nuclear generation, but with some loss of efficiency in electricity generation.

Possible options for the 2032 generation mix are shown below. The KA CARE plan has been taken as a base, even though it is rather out of date, because of its reasonable level of detail, and its inclusion of nuclear power. Some gas turbines have been added to the KA CARE plan to assure that peak summer daytime demand plus the National Transformation Plan's stated 12% reserve margin can be met. Figures here assume conservatively that no wind power can be counted as firm capacity, that the GCC interconnection is not expanded, that no additional pumped hydro storage nor other storage is added, and that the Egypt interconnector is not usable to meet peak

---

432. Nachet, S. & Aoun, M. C., 2015. *The Saudi electricity sector: pressing issues and challenges*. [Online] Available at: https://www.ifri.org/sites/default/files/atoms/files/note_arabie_saoudite_vf.pdf
433. AUB Policy Institute, 2017. *Solar vs Nuclear: Which is cheaper for Desalination?*. [Online] Available at: https://www.aub.edu.lb/ifi/Documents/publications/policy_briefs/2016-2017/20170215_solar_nuclear.pdf

demand.[434] Coal is not considered as an option due to its polluting nature, the lack of domestic or indeed regional resources, and the energy security consequences of importing it (although some studies have been carried out, there are no plans for coal use in the country).

Only CO2 emissions from fuel combustion are considered, not life-cycle emissions for construction and fuel production. Waste-to-energy is assumed to be carbon-neutral (its contribution in any case is small and it does not vary across the scenarios considered).

The options are constructed so that daytime and night-time summer and winter demand can be met with a minimum 12% margin, and that total generation over the year is sufficient to meet total demand. Assumed maximum annual capacity factors are 90% for all thermal, nuclear, waste-to-energy and geothermal generation; 40% for solar CSP; 20% for solar PV; and 30% for wind. CSP is assumed to have 8 hours of storage so that it can meet the early evening demand peak.

The options considered are as follows:

1) KA CARE plan as described above

2) KA CARE plan but with 18 GW of combined-cycle gas turbines replacing the planned 18 GW of nuclear

3) KA CARE plan but with an additional 21 GW of CSP (taking the total to 46 GW) replacing the planned 18 GW of nuclear

4) KA CARE plan but with no nuclear, more CSP as in option (3) and with solar PV increased from 16 GW to 74 GW

5) KA CARE plan with 18 GW of nuclear and with solar PV increased from 16 GW to 64 GW of solar PV (this saturates winter daytime demand)

---

434. This is on the grounds that, as the sun sets in western Saudi Arabia and solar output falls, Egypt would be entering its own peak evening demand period.

6) Current system expanded to meet demand with only combined cycle oil/gas generation and currently-tendered renewables

7) Maximum build-out of nuclear to meet the minimum load (winter night-time) with CCGT meeting the rest

Capacity (GW)	2015 fleet[435]	2032 fleet						
		(1) KA CARE	(2) KA CARE, CCGT replaces nuclear	(3) KA CARE, CSP replaces nuclear	(4) KA CARE, CSP replaces nuclear; more PV	(5) KA CARE; more PV	(6) Current system expanded	(7) High nuclear
Steam turbine	32	0	0	0	0	0	32	0
Gas turbine	38	21.64	21.64	18.64	10.34	13.33	38	23.8
Diesel generator	1	0	0	0	0	0	1	0
Combined-cycle turbine	10	60[436]	78	60	60	60	73.2	68
Solar PV	0.085	16	16	16	74	64	0.3	0.3
Solar CSP	0	25	25	46	46	25	0.111	0.111
Wind	0.00275	9	9	9	9	9	0.4	0.4
Geothermal	0	1	1	1	1	1	0	0
Waste-to-energy	0	3	3	3	3	3	0	0
Pumped hydro	0	2[437]	2	2	2	2	2	2
Nuclear	0	18	0	0	0	18	0	52
GCC Interconnection	1.2	1.2	1.2	1.2	1.2	1.2	1.2	1.2

These scenarios are, of course, simplistic and ignore issues such as the finer scale of demand, seasonal correlations between wind and solar generation, the geographic location of generation and demand,

435. ECRA, 2015. *Statistical Booklets.* [Online] Available at: https://ecra.gov.sa/en-us/MediaCenter/DocLib2/Pages/SubCategoryList.aspx?categoryID=5
436. Assumed that all thermal generation in KA CARE plan would be combined-cycle burning either gas or diesel
437. Not explicitly in the KA CARE plan, but under construction today

and the time for plants to ramp up and down. They are conservative in ignoring options for storage (other than the planned pumped hydro and CSP), some level of firm wind capacity, new interconnections, or demand response. They also exclude other energy demand than electricity which could also be met by nuclear or renewables, for instance industrial process heat.

The generation mixes are shown graphically in Figure 11, along with the peak summer daytime and night- time demand (plus the desired 12% reserve margin). Demand is shown with a business-as-usual case, and a case with Oliver Wyman's projected 27% decrease in demand through efficiency measures by 2032. It can be seen that the efficiency measures save about 40 GW of peak summer daytime required capacity and 33 GW of peak summer night-time required capacity. The two cases with large amounts of solar PV (4 and 5) have higher overall capacities because of the lower capacity factor of solar and the requirement for back-up generation at night.

Figure 11 Generation mixes of different options

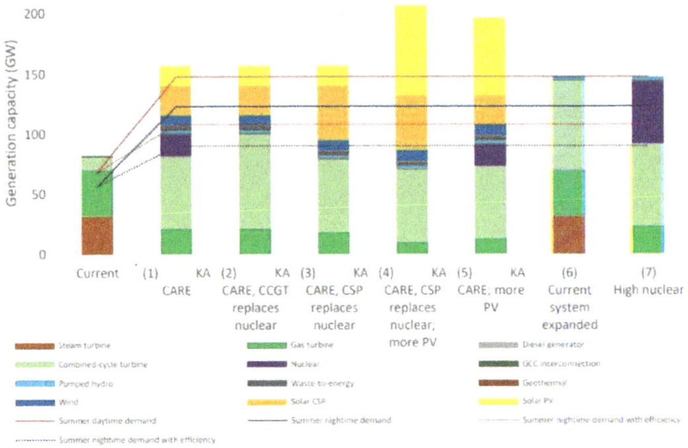

Average cost per year of the electricity generated (fuel, operating and capital cost amortisation) is shown below, along with the average gas consumption. Note that costs exclude transmission and distribution. The fuel price assumed is $7 per MMBtu as a reasonable estimate of long-term LNG import prices or the cost of producing domestic unconventional gas. This assumes the Saudi gas market is reformed so the marginal gas supply sets the price. In this case, industrial gas demand would probably fall (or at least not rise so quickly), making more gas available. Electricity tariffs would have to be increased to avoid incurring a large subsidy burden, and this would also restrain demand. However, these additional effects have been ignored at this stage for the simplicity of comparison.

If instead Aramco's average cost of supply is used, this would be significantly lower given the contribution of low-cost associated gas as well as large non-associated conventional fields. This would make gas-fired generation appear cheaper, but at the cost of a (hidden) subsidy from Aramco to the power sector. If Aramco were able to reduce unconventional gas production costs significantly, the overall cost of a given generation mix would be lower, but it would probably not change the balance much given that wind (in good locations) and solar PV are likely still to be cheaper, while nuclear is still more expensive. However, it would be less favourable for CSP.

Scenario	(1) KA CARE	(2) KA CARE, CCGT replaces nuclear	(3) KA CARE, CSP replaces nuclear	(4) KA CARE, CSP replaces nuclear; more PV	(5) KA CARE; more PV	(6) Current system expanded	(7) High nuclear
Average gas consumption (Bcf/day)	7.55	10.2	8.83	6.93	5.98	14.93	5.69
Annual cost ($ billion)	62.1	60.4	61.5	56.5	58.0	58.3	61.7
Average cost of electricity	8.67	8.43	8.58	7.88	8.09	8.14	8.62
Supplied (₡/kWh)[438]							
Annual CO2 emissions (million tonnes)	153.3	207.1	179.2	140.6	121.3	303.9	115.8

Gas consumption is higher in all cases than current levels, since oil burning for power is assumed to be completely phased out. However, from all the options 1-5, the highest estimated gas use, in option 2, is equivalent to about 4 million TJ, while total 2014 consumption of natural gas, crude and oil products for power generation was about 3.6 million TJ. Therefore, even the most fossil-fuel intensive option only has a moderate increase in consumption while doubling generation. By contrast, option 6, relying on the current path of fossil-fuelled generation, has much higher gas (and/or oil) consumption at about 5.75 million TJ.

It is not clear what proportion of the planned increase in natural gas production is intended to be devoted to the power sector. 2014 marketed production was 9.9 Bcf/day (on OPEC/BP figures; as noted, the IEA is lower), of which 4.6 Bcf/day goes to power generation, or 47% (as noted, on the IEA's figures, the share is 56%). An increase to 23 Bcf/day gross gas production by 2026, if then held constant to 2032, with an implied 20 Bcf/day of marketed gas, would mean that in the above options 1-5, power generation would consume 30-51% of available gas, while all current oil consumption

438. Excluding transmission & distribution costs

for power would be freed up for export or other uses. Available gas for industry or other uses would increase from about 5 Bcf/day in 2014 to 10-14 Bcf/day in 2032, which seems sufficient to meet substantial growth. Therefore, all these scenarios seem reasonable in terms of gas use, assuming the production targets are met, or domestic production is supplemented with imports. By contrast, option 6 would consume 75% of marketed gas, leaving little for industry.

If gas production only remained constant at current levels, then it could be supplemented with oil for power. Combined-cycle gas turbines would in principle have to run on clean fuels (diesel), which is significantly more expensive. However, the total quantities of liquid fuels burnt in options 1-5 would be less than today's quantities except in option 2 where they would be slightly larger and 3 where they would be similar. In the fossil-fuelled current path with gas production at today's levels, oil burning would reach about 1.8 million barrels per day.

It can be seen that the five renewable/nuclear options are all rather similar in cost, since the estimates used here for CCGT (at $7 per MMBtu gas), CSP, and nuclear are very close. However, adding additional PV does bring down the total cost by saving gas, given its very low-costs. Adding more wind, which would also contribute at night, would also reduce gas consumption and total costs. These options are vulnerable to CSP costs, given the uncertainty in its price levels and that it is relied on to contribute to meeting summer evening peak demand. If it turned out to be significantly more expensive, then more gas-fired and/or nuclear generation would be required. Arguably about 6-7 GW of CSP is possible at low-cost by hybridisation with combined-cycle gas plants, if suitable land is available near the plants. The fossil- fuelled option (6) is also similar in cost, assuming that the required volumes of fuel are available. The nuclear-heavy option (7) has a somewhat greater cost given the higher (but still probably over-optimistic) cost of nuclear assumed.

All the renewable/nuclear plans represent substantial reductions in CO_2 emissions compared to the estimated 232 million tonnes of electricity-related emissions in 2014, despite much higher total generation. This is achieved by switching entirely to natural gas instead of oil, using more efficient combined-cycle gas turbines almost exclusively, and introducing low/zero-carbon generation. Savings range from 25 to 112 million tonnes of CO_2 per year versus 2014, while the high nuclear (7) produces slightly less CO_2 than the most aggressive renewables option. The fossil-fuelled option (6) sees a substantial rise in CO_2 emissions, despite the use of high-efficiency CCGTs and the (assumed) replacement of all oil-fuelled generation with gas. If insufficient gas were available, the use of oil would raise emissions further.

Replacing nuclear with CCGT or solar CSP raises CO_2 emissions significantly (since solar CSP has a lower assumed load factor and hence the CCGTs run more). However, the option with much more solar PV is not only cheaper overall but has somewhat lower emissions than the KA CARE case with nuclear. Lowest emissions of all are achieved in the case with both nuclear and increased PV.

All these options would also greatly reduce most other pollutants, notably sulphur oxides from heavy fuel oil, assuming that the primary fuel is gas.

The pace of generation construction in the options presented amounts to 3.6-8.9 GW of renewables per year during 2019-2032, and up to 1.8 GW of nuclear per year (assuming a 2022 start) in the KA CARE case. This would be a significant acceleration on the current tenders for 1.7 GW of renewables. Solar capacity added in Germany per year reached a maximum of about 7 GW between

2010-12.[439] The UAE began constructing its 1.4 GW reactors in 2012 and the first is due online now likely in 2021. The high nuclear option (7) would require 52 GW of nuclear to be installed in 10 years (France has 63 GW total installed), a pace comparable to installing almost the UAE's entire nuclear programme (5.6 GW) every year from 2022 to 2032. This appears completely implausible, and therefore the option is shown only for illustration.

Therefore while the renewables and the lower nuclear installation rates appear plausible, they are challenging and would require a quick start with a minimum of bureaucratic delay. More reliance on local content would also slow the programme. However, both renewables and nuclear appear to face similar challenges in terms of installation pace.

All these options assume a quite rapid conversion of current generating plant to combined-cycle operation and/or the construction of new CCGTs. The current fleet of 32 GW of steam turbines, 38 GW of gas turbines, and 10 GW of CCGT has to be converted to 60-78 GW of high-efficiency modern CCGT by

2032 plus 10.3-21.6 GW of gas turbines for peaking use. If older plants are retained in service or not converted, the efficiency would be lower and hence more oil/gas fuel would be required. This study has not assessed how much of the existing fleet would be suitable for conversion to CCGT. However, even if all the additional CCGT capacity had to be newly-built, the rate of 3.5 GW per year is still feasible as thermal generation capacity has been rising at about this rate historically.

---

439. Energie Wende: Umschalten auf Zukunft , 2019. *Time Series for the Development of Renewable Energies in Germany.* [Online] Available at: https://www.erneuerbare-energien.de/EE/Navigation/DE/Service/Erneuerbare_Energien_in_Zahlen/Zeitreihen/zeitreihen.html

5.1.3. Efficiency

Energy use in Saudi Arabia is highly inefficient by global standards. At 6.9 tonnes of oil-equivalent per capita in 2014, the 10th-highest in the world, it was slightly ahead of the U.S. (6.9 toe/capita) and well ahead of the EU (3.2 toe/capita) and the world average (1.9 toe/capita). As opposed to most countries, whose energy intensity has been falling, Saudi energy intensity has been rising over the long term and reached 5.8 MJ per $ of GDP (2011 dollars, purchasing power parity) in 2015, ahead of 4 MJ per $ of GDP in the EU.[440]

Saudi Arabia has launched the Saudi Energy Efficiency Center to develop the Saudi Energy Efficiency Policy, an inter-agency effort. This is focused on industry, buildings, and transport, primarily by means of benchmarking and standards. Studies have been conducted for a national District Cooling Company and a national Energy Services Company (ESCO). The programme's overall aim is to increase electricity efficiency by 30% by 2030. Building codes have been strengthened, though this will take a long time to have an impact on the existing stock, and monitoring and compliance is likely to be lacking initially.

Areas for gains in efficiency include air-conditioning, in particular, as by far the biggest user of electricity. District cooling use is low in Saudi Arabia, at less than 5%, versus 30% in Dubai.[441] More efficient air-conditioners, better maintenance, and improved insulation are being targeted (70% of existing buildings are not insulated).[442] Other promising areas include public transport such as

---

440. Nachet, S. & Aoun, M. C., 2015. *The Saudi electricity sector: pressing issues and challenges.* [Online] Available at: https://www.ifri.org/sites/default/files/atoms/files/note_arabie_saoudite_vf.pdf
441. Global Market Insights, 2019. *Middle East District Cooling Market.* [Online] Available at: https://www.gminsights.com/industry-analysis/middle-east-district-cooling-market
442. IEA, 2015. *Energy Efficiency Standards and Mandates: Saudi Arabia.* [Online] Available at:

the Riyadh Metro, under construction; appliance and vehicle efficiency standards; more efficient lighting; solar water heating; and heat pumps, which could drive absorption chillers as well as providing space- and water heating.

One KAPSARC study suggests that a building retrofit programme could save 3.7-22.9 GW of peak electricity capacity, depending on the level of retrofit, with reasonably attractive payback periods, depending on the electricity price,[443] though the payback periods are rather long for the higher levels of ambition.

An Oliver Wyman study[444] concluded that efficiency could cut 27% of electricity demand from the 2030 baseline (with technical feasibility up to 56%) at an assumed electricity price of 10 ₡/kWh (comparable to the numbers calculated above, particularly when transmission and distribution costs are added). If the drop in peak demand by 2032 is proportionate, then it would be reduced from 132 GW to 96 GW; i.e. the annual growth rate would fall from 5% to 2.5%.

Such efficiency gains would substantially improve the feasibility of meeting Saudi Arabia's electricity demand without nuclear power, as they would reduce the required pace of renewable and/or CCGT construction, and save on fuel demand and CO2 emissions. Achieving efficiency gains is, though, very dependent on continuing to reform and reduce subsidies, as the still low electricity and water prices do not encourage energy-saving behaviour or the use of

---

https://www.iea.org/policiesandmeasures/pams/saudiarabia/name-147402-en.php
443. Dubey, K., Howarth, N. & Krarti, M., 2016. *Evaluating Building Energy Efficiency Investment Options for Saudi Arabia,* Riyadh: KAPSARC.
444. Oliver Wyman, 2013. *Delivering on the Energy Efficiency Promise in the Middle East.* [Online] Available at:
https://www.oliverwyman.com/content/dam/oliver-wyman/global/en/files/archive/2013/Energy_Efficiency_inMiddleEastFINAL.pdf

efficient construction and equipment, and replacing them with a more targeted form of social support.

5.2. Solar

Saudi Arabia has among the best conditions for solar power in the world. Global horizontal irradiance (relevant for photovoltaic power) in Riyadh is 2242 kWh/m2 per year (see Figure 12), while direct normal irradiance (relevant for concentrating solar thermal power, CSP) is 2116 kWh/m2 per year (see Figure 13). For comparison, San Diego has 1964 kWh/m2 and 2242 kWh/m2 respectively. The western parts of Saudi Arabia are superior for CSP, as the climate along the Gulf can be hazy and dusty.

Figure 12 Global horizontal irradiation[445]

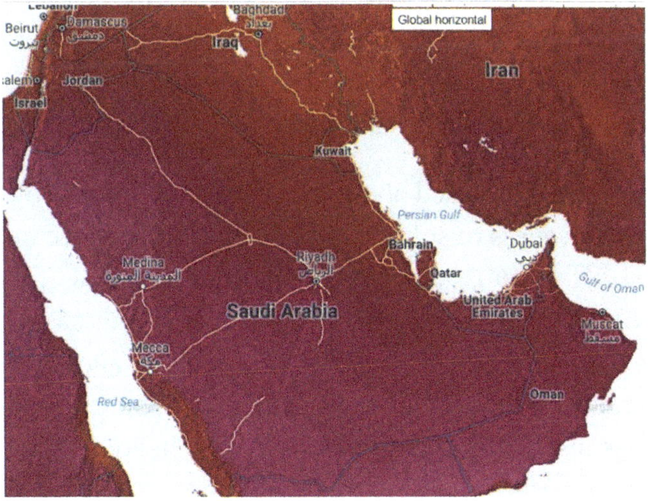

445. Solargis, 2019. *iMaps.* [Online] Available at: https://solargis.info/imaps/#c%3D25.918526%2C41.813962%26z%3D5

Figure 13 Direct normal irradiation[446]

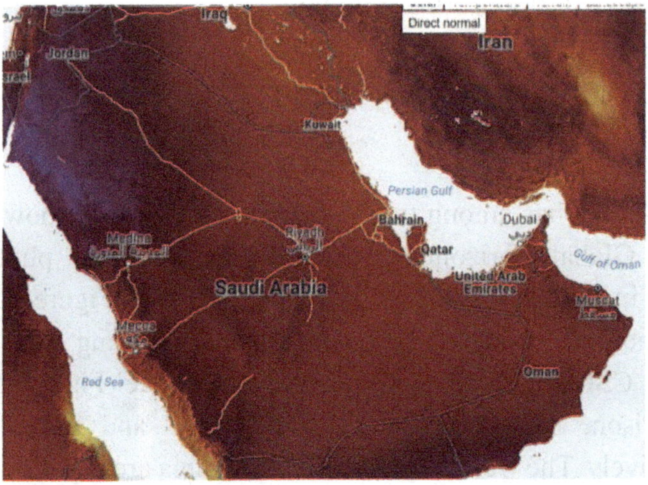

As can be seen from the maps, conditions for solar PV in Saudi Arabia are at least as good as in the UAE, which has launched several very successful and cost-competitive solar projects to date. Conditions in the west of the country actually appear to be superior. Conditions for CSP are similar to those in the UAE, except in the north-west where they are superior.

Concerns are often raised about the impact of dust and blown sand on the effectiveness of solar PV in the Gulf. Experience in the UAE suggests that these issues are manageable with a careful cleaning schedule and do not add significantly to operating costs.[447] Given low population densities in much of the country, and large tracts of arid land with little alternative use, finding suitable cheap land for large solar installations should not be problematic, although some transmission costs may have to be incurred.

---

446. Solargis, 2019. *iMaps.* [Online] Available at: https://solargis.info/imaps/#c%3D25.918526%2C41.813962%26z%3D5
447. First Solar and Masdar personnel, personal communication

The Saudi population is concentrated on the east coast (around the oil and industrial centres of Dammam, Dhahran and Jubail), the centrally-located capital Riyadh, and the west coast with the holy cities of Makkah and Madina, the port of Jiddah, and industrial city of Yanbu' (see Figure 14 – this is for the year 2000 so the population has grown substantially but the overall pattern remains similar). Current demand is roughly evenly split between east, centre and west (~16 GW peak demand each) with a remaining ~10 GW in the south and Yanbu' Industrial City.[448]

Figure 14 Saudi Arabia population density (2000)[449]

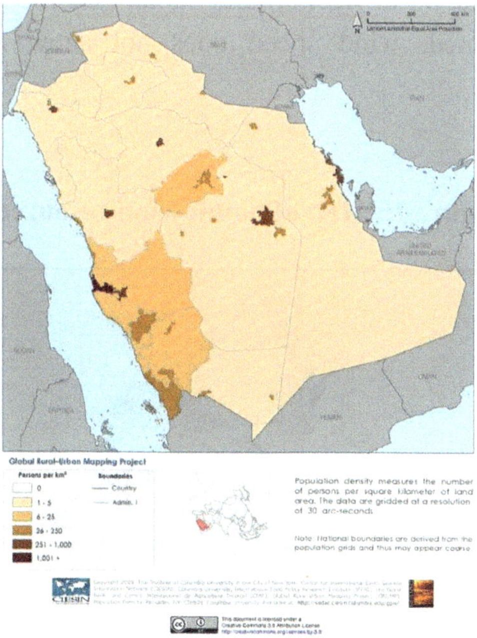

448. Ghabban, A. A., 2013. *Saudi Arabia's Renewable Energy Strategy and Solar Energy Deployment Roadmap,* Riyadh: KACARE.
449. SEDAC. *Population Density Grid, v1 (1990, 1995, 2000) » Maps.* [Online] Available at: https://sedac.ciesin.columbia.edu/data/set/grump-v1-population-density/maps/6

The difference in sunset times between Dammam on the east coast and the industrial city of Yanbu' on the west coast is about 40 minutes in summer. There is thus some scope for solar generation in the west to serve the east's early-evening peak demand. The large concentration of population in the west would, however, require another solution. This could be via the new interconnector to Egypt, if surplus capacity is available there at the right time.

5.3. Wind and Other Renewables

The best Saudi wind resource is located on the west (Red Sea) coast, especially in the north-west around Tabuk (see Figure 15). In most sites investigated, wind speed does not show much monthly variation.[450]

Figure 15 Map of Saudi wind resource[451]

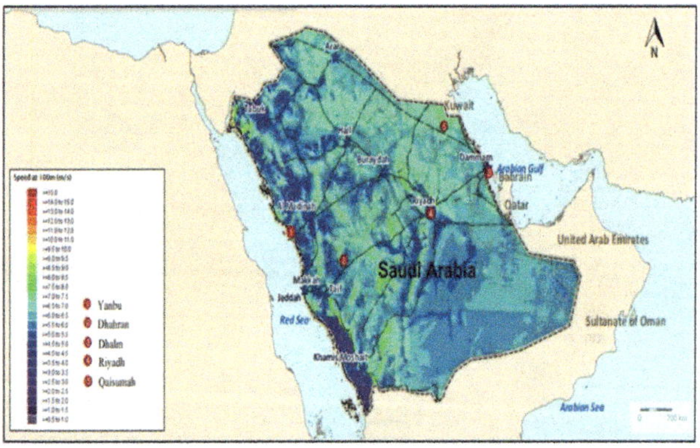

---

450. Krishna, L. V. & Thalhi, F. A. A., 2015. Solar and Wind Energy Potential in the Tabuk Region, Saudi Arabia. *International Journal of Applied Science and Technology* , 5(3), pp. 12-22.
451. Mohamed, M. A., Eltamaly, A. M. & Alolah, A., 2016. PSO-Based Smart Grid Application for Sizing and Optimization of Hybrid Renewable Energy Systems. *PLOS One,* Volume 11, Figure 1.

It is assumed here that the 9 GW target given by KA CARE is feasible in terms of total capacity of suitably windy sites. The initial bids for the Dumat Al Jandal site, which is in north-central Saudi Arabia near the Iraqi and Jordanian borders, are very promising in terms of low-costs (the Sakaka solar PV plant is in the same area). No information was identified as to the ultimate wind power potential in Saudi Arabia. With relatively low wind speeds over most of the country, larger amounts of wind power may be possible but with higher costs.

Geothermal potential is present around Jizan in the south-west, where there are hot springs,[452] and in various granites in the west of the country with high heat-flow.[453] Lower-temperature resources may be useful for thermally-driven cooling or desalination. Estimated potential generation is 3-5 GW and hence it is assumed here that the KA CARE target of 1 GW by 2032 is feasible.

Different studies estimate waste-to-energy potential by 2032 of 1.447-2.074 GW.[454] However, a recycling policy would substantially reduce this potential (though it would be better for life-cycle emissions). The KA CARE target of 3 GW may therefore be somewhat ambitious and up to 1-3 GW of other generation may be required instead. However, the KA CARE target has been used here as the impact is relatively small and it does not vary between scenarios.

---

452. Richter, A., 2017. *Saudi Arabia eyeing geothermal energy for future energy mix.* [Online] Available at: http://www.thinkgeoenergy.com/saudi-arabia-eyeing-geothermal-energy-for-future-energy-mix/
453. Proceedings World Geothermal Congress , 2015. *Geothermal Energy Resources of Saudi Arabia: Country Update.* [Online] Available at: http://www.bu.edu.eg/portal/uploads/Science/Geology/3501/publications/Aref%20Adel%20Ahmed%20Elaraby%20Lashin_01014.pdf
454. http://www.sciencedirect.com/science/article/pii/S1364032116300223, Ouda, O.K.M. 'Assessment of Waste-to-Energy Potential in Saudi Arabia as Electricity Source & Environmental Protection Measure' Prince Mohamed Bin Fahd University

5.4. Energy Storage

Saudi Arabia is currently constructing two 1000 MW pumped storage facilities on the west (Red Sea) coast, one using seawater and the other desalinated water,[455] with storage of 8 hours (implying 8000 MWh). Additional sites have been identified and so more storage could probably be installed if required.[456]

As noted, integrated thermal storage was also assumed for the CSP plants included in the generation mix above.

Other storage options, such as batteries, have not been evaluated here, but could be useful in balancing demand and renewable output. Particularly appropriate to the Saudi climate is thermal storage, mainly for air-conditioning given its large share of load, which would be more effective if building insulation is improved. Thermal storage can include pre-cooling buildings before the hottest part of the day or making integrated with district cooling plants, which are anyway inherently more efficient than individual cooling units. Another option is to use reverse osmosis plants for desalination in low-demand periods, and store the water, as noted above.

---

455. AF Consult, 2015. *ÅF is awarded a new contract for consulting services for Pumped Storage Hydro Power Plant in Saudi Arabia.* [Online] Available at: https://www.afconsult.com/en/newsroom/press-releases/all/2015/af-is-awarded-a-new-contract-for-consulting-services-for-pumped-storage-hydro-power-plant-in-saudi-arabia/; Elamin, G., Shwehdi, M. & AlAwaji, S., 2015. Assessing the Electricity Storage for Renewable Energy in Saudi Arabia. *International Journal of Computer Science and Electronics Engineering (IJCSEE),* 3(5).
456. SNC Lavalin, 2013. *Pre-Feasibility Study of a 1000 MW Pumped Storage Plant in Saudi Arabia.* [Online] Available at: https://www.slideshare.net/SNC-Lavalin/130715-hydrovision-paper-v3

6. Security

Despite some terrorist incidents and a recent reported drone attack on oil pumping stations, the physical security of Saudi Arabia's oil installations has been assured by extensive defensive measures. There is therefore no unusual security risk to relying on domestic oil and gas output for energy, except in case of a total breakdown in national security (which would also badly affect demand). Large, dispersed renewable energy sites do not present a single point of failure and hence would also not be particularly vulnerable to security threats, although long-distance transmission lines could be threatened.

Any strategy depending on imports of natural gas, LNG (or coal) could be vulnerable to various supply disruptions such as boycotts, sanctions, unrest in exporting countries, or threats to maritime supply routes through the Gulf and Red Sea.

Nuclear power plants would, like the oil facilities, likely be highly fortified and built to withstand some level of plausible attacks or natural disasters. Of the sites proposed so far, Jizan is near Yemen and may be problematic given the continuing war there and some cross-border incursions and reported missile attacks. Jubail is in the Eastern Province, with a large Shi'ite Muslim population that has often complained of discrimination and where unrest is frequent. It is also relatively close to Iran, with whom Saudi Arabia has poor relations at present. The proposed site at Tabuk in the north-west, is relatively close to Israel, which presumably would be expected to agree tacitly to a Saudi civil nuclear endeavour, but otherwise does not present any obvious security concerns. The two most recently short-listed sites on the Gulf (and so relatively near Iran) are close to Qatar, which is still under Saudi boycott, but does not pose a military threat.

7. *Conclusions*

Saudi Arabia's fast-growing energy demand and heavy reliance on oil and gas for power generation have led to investigations of nuclear power as one component of its future energy mix. Nuclear power has been under consideration for a long period, with frequent changes of plan. Both nuclear and renewable energy have made little headway, but several tenders for solar and wind projects have recently been launched.

This study has assessed options for meeting Saudi Arabia's future energy mix, primarily considering electricity. Based on the original KA CARE plan, a number of plans with varying mixes of renewables, nuclear and fossil fuels were considered. These scenarios suggest that the 2032 demand can be met with a mix of gas-fired (highly efficient CCGT) and renewable energy at a cost comparable to, indeed slightly lower than, options including nuclear. Proposed renewable build-out rates are ambitious but should be achievable. The proposed renewable mix is primarily solar (CSP and PV), for which Saudi Arabia has excellent resources. Both the renewable and nuclear mixes deliver substantial declines in CO2 emissions, with most oil eliminated from the power sector, and moderate increases in gas demand that should be achievable within Aramco's gas production targets.

These options do not assume significant gains in energy efficiency, which would further reduce costs, CO2 emissions, and required construction pace, nor do they optimise energy storage or allow for continuing cost reductions in renewable (or nuclear) energy.

Nuclear power may have other strategic, technological, or development-related attractions for Saudi Arabia. However, overall, it appears that it would be economically, practically, environmentally, and organisationally feasible for it to meet its energy needs without resorting to nuclear power.

Meeting Iran's Energy Requirements without Nuclear Power

Patrick Clawson

This study is about how Iran can meet its electricity requirements without nuclear power. To be clear, let us explain what this study is not about. Most of Iran's energy needs are met with natural gas and oil, which is used for transportation, space heating, and the like. Nuclear power is of little relevance to meeting those needs. Where nuclear power comes into play is in Iran's generation of electricity. Therefore this paper is only about supplying Iran's electricity without nuclear power.

This study is also not about the proliferation risks posed by Iran's nuclear choices, be that for power plants, a full nuclear fuel cycle, or various research programs. While the effectiveness of various international agreements and the potential for Iran to weaponize its nuclear capabilities are extremely important questions, they are not the subject of this paper. Instead, this study examines to what extent Iran has a legitimate civilian purpose for nuclear power, which is in many ways a precursor to those weapons issues. Bear in mind that its nuclear power program is the principal explanation Iran has given for its robust nuclear activities (while Iran also cites medical, agricultural, and scientific purposes, those are distinctly secondary to the power program).

This paper examines where nuclear power fits in Iran's electricity situation. The paper begins by examining the current electricity generation situation in Iran and the likely prospects if there is no change in policies – prospects which include additional nuclear power plants. The bulk of the paper considers policies more appropriate than nuclear power – policies which would save Iran vast sums while at the same time offering environmental results as good or better than nuclear power. The policies considered include, in order of their potential savings to Iran: cutting subsidies and therefore electricity demand; relying on more renewable energy; replacing much single-cycle gas generation with combined-cycle; and relying on natural gas for much additional capacity. The paper concludes with a discussion of institutional factors which contribute to the decision to pursue additional nuclear power plants.

The Current Electricity Generation Situation

Given the many clandestine activities of the Iranian government (e.g., support for terrorism and subversion), its history of misleading international inspectors about its nuclear program, its record of election manipulation (most blatantly in the 2009 presidential election), and its authoritarian government with only a democratic veneer, it is not surprising that many Westerners assume that little information is available from the Iranian government. In fact, quite the opposite is true. On the whole, the Iranian government provides more data on the internet about its activities than the U.S. government does (with obvious exceptions about national security). In particular, the data available about electricity in Iran is exhaustive – and much of it is in English.

In 2015 – meaning the Iranian year which ends on March 20 of the next Western calendar year – Iran generated 280 billion kilowatt-hours (KwH) of electricity.[457] Of that total, 123 billion KwH came

457. "Detailed Statistics of Iran's Electricity Industry - 1395 [2016/17]" in Persian, MOE/Tavanir, page 1, http://isn.moe.gov.ir/getattachment/05b65404-

from facilities owned by the Ministry of Energy (MOE); note that MOE is largely concerned with electricity, since the separate Ministry of Oil is responsible for all aspects of oil and natural gas production and distribution (see Table 1.21, Gross Power Generation). Of the non-MOE electricity, 153 billion KwH came from facilities owned in theory by the "private sector" – which is not very private -- and connected to the national grid, 6 billion KwH from facilities connected to and owned by large industries, and less than 1 billion KwH from other facilities not connected to the national grid – those being in isolated regions.

Even more so than in most countries, the "private sector" facilities are highly dependent on government policies, as discussed below. These facilities are a recent phenomenon. The first non-MOE generation was only in 2000.[458] As late as 2010, non-MOE facilities were only 17% of total operational capacity; 2015 was the first year in which non-MOE operational capacity exceeded that of MOE (see Chart 1.1, Operational Capacity of Power Plants by Ownership). Most of the growth of private facilities has come from privatization of facilities built by MOE, not new construction. Consider that between 2010 and 2015, the capacity of non-MOE grew 23,596 MW while total national capacity – MOE plus private – grew 10,639 MW.

Gas has long predominated in the Iranian power-generation mix. At the time of the 1978 revolution, Iran's capacity was 6,362 MW, of which hydro was 1,804; diesel 447; and gas 4,111, with 1,606 steam and 2,505 single-cycle gas turbine (see Table 1.2, Operational Capacity of Power Plants, and Chart 1.2, Operational Capacity of Power Plants by Fuel Type). In recent years, Iran is increasingly

---

b61f-4dd3-9490-2ef35c6a965f/امار-تفضيلي-صنعت-برق-ايران،-مديريت-ويژه-راهبردي،-
س. This very lengthy report has, as the title indicates, detailed statistics.

458. All of the data in the rest of this section, unless otherwise noted, are from the MOE/Tavanir *Statistical Report on 29 Years of Activity of Iran Electric Power Industry (1967-2015)* in English, .

moving towards combined-cycle technology. The first combined-cycle facilities became operational in 1997, and by 2015, such facilities made up 23% of national operational capacity. In 2015, the operational capacity of Iran's power plants was 64,708 megawatts (MW), of which 61,992 used natural gas, 284 used diesel, 11,278 were hydro facilities, 1,140 were nuclear, and 53 were renewable. MOE owned 32,119 MW; others, 32,589 MW. Of the natural gas facilities, 15,210 were steam (10,942 MOE and 4,268 non-MOE); 21,630 single-cycle gas turbine (5,086 MOE and 16,544 non-MOE), and 15,112 combined cycle (3,389 MOE and 11,723 non-MOE).

Iran has been rapidly expanding its operational capacity for many years. From 1967 to 1978 – the Shah's last year -- operational capacity rose 650%, an average annual increase of 18%. More recently, capacity has almost tripled since the first combined-cycle facilities started about twenty years ago, going from 23,050 in 1996 to 64,708 in 2015, an average annual increase of 5.58%.

Whereas the additional capacity under the Shah was mostly single-cycle gas turbine, the revolutionary government in its first decades mostly built steam plants. Of the 14,848 increase in operational capacity from 1978 to 1996, 9,850 was steam and 4,874 was single-cycle gas turbine; there was little change in hydro or diesel. More recently, the pattern has shifted, with a burst of hydro dam construction, considerable combined-cycle gas, and continued considerable addition of single-cycle gas turbine. In the period 1996-2015, of the 41,658 MW capacity increase, 15,112 has been combined cycle; 14,251 single-cycle gas turbine; 3,754 steam; 9,279 hydro; 1,140 nuclear; and 53 renewable, while diesel capacity shrank 194.

Of the fuel consumed for electricity generation, gas predominates. In 2015, gas provided 525.81 million tons of oil equivalent (MTOE), while fuel oil provided 5.85 MTOE (see Table 1.3, Fuel Consumption of Power Plants). Note that because gas supply is insufficient to meet all demand during the peak consumption period in the winter, gas is provided for heating and industrial use while electricity generators have to shift to using fuel oil. One Iranian study comparing the economics of differing fuel sources used as an assumption that fuel oil would be burned three months a year.[459] From the Iranian data, it would appear that the shift takes place primarily at MOE facilities; the non-MOE facilities use very little fuel oil or diesel, even though some of the operational capacity of such facilities is steam, which can handle the fuel switch over more readily than can turbine facilities. That said, even the most recently built gas-fired turbine generating facilities have to include large storage facilities for fuel oil, and news reports treat as a matter of fact that fuel oil will be burned in the winter. The Iranian data show that the problem became worse after 2009 as natural gas use for electricity generation actually fell (see Chart 1.3, Trend of Power Generation of Thermal and Non-Thermal Power Plants by Fuel Types, taken directly from an Iranian report).[460]

Electricity is transported and distributed to consumers nearly entirely by one nationalized company, Tavanir, which is effectively part of MOE. There do not appear to be any plans to change that situation.

A brief word is in order about the place of electricity in Iran's overall energy scene. Iran is a heavy consumer of energy. In 2016, according to the widely-used BP *Statistical Review of World Energy,*[461] an annual report with accompanying spreadsheet showing historical data going back many decades, Iran's primary energy consumption

---

459. M. Khakbazan-Fard et al., "Generation Expansion Planning for Iran Power Grid," *Iranian Journal of Science and Technology – Transaction B: Engineering*, Vol 34. No. B3, 2010.
460. Note that the Iranian data use convert the fuel types with slightly different conversion factors than the BP factors used elsewhere in this study.
461..*Statistical Review of World Energy*, British Petroleum, June 2017, http://www.bp.com/en/global/corporate/energy-economics/statistical-review-of-world-energy.html

(271 million tons of oil equivalent) was nearly twice that of Turkey's (138 million tons of oil equivalent), a country of similar population and larger national income. Such heavy use of energy is common among the Gulf oil producers. In 2016, primary energy consumption per capita was 3.37 tons of oil equivalent in Iran, 8.26 in Saudi Arabia, 10.19 in Kuwait, and 12.28 in the United Arab Emirates – compared to 3.21 in the European Union and 7.03 in the United States.[462] Iran and the Gulf oil producers countries suffer from many of the same wasteful energy policies, be it regarding electricity or such matters as subsidized gasoline.

Likely Prospects Without a Change in Government Policy

The policies presently in place are likely to lead to robust growth in electricity demand. While nearly all increased demand will be met by natural gas, Iranian authorities are determined to add additional nuclear power plants, though these plants will provide only a small portion of the additional capacity. Part of the reason that these plans are likely to come to pass is the sad record of past reform efforts.

Current Plans

As part of the Stanford Iran 2040 Project, in April 2017, four scholars presented a paper entitled "The Outlook for Natural Gas, Electricity and Renewable Energy in Iran."[463] They argue that

462. Author's calculations based on energy consumption data from the BP source cited above and population data from the World Bank.
463. Pooya Azadi (Stanford), Arash Nezam Sarmadi (National Grid), Ali Mahmoudzadeh (U British Columbia), and Tara Shirvani (World Bank), "The Outlook for Natural Gas, Electricity, and Renewable Energy in Iran," Stanford Working Paper No. 3, April 2017, https://iranian-studies.stanford.edu/iran-2040-project/publications/outlook-natural-gas-electricity-and-renewable-energy-iran.

through 2040, Iran is likely to add capacity at the rate of 1,300 MW per annum, compared to 3,000 MW per annum in the recent past (in actuality, operational capacity increased on average 2,400 MW per annum in the period 2006-2015). Given that Iran's operational capacity in 2015 was 64,708 MW, it would imply a growth rate of 1.6% per annum in the period 2015-2040. Other estimates are much higher. A late 2015 estimate by the respected private firm BMI Research forecast that electricity generation would grow at an average 2.5% per annum from 2015 through 2024.[464] At the 2013 9th International Energy in Tehran, Hamed Shakouri and Allyah Kazemi from the University of Tehran presented a complex model that forecast 6.3% annual growth in Iranian electricity demand for the foreseeable future.[465] The Iranian government's forecast growth rate is higher than any of these outside predictions. Iran's Sixth Five-Year Development Plan, covering 2017-21 and approved by the Majlis (parliament) in April 2017, calls for adding 25,000 MW over the five years, which would be an average annual rate growth rate of 6.75%. [466]

A major driver of investment decisions is peak consumption, rather than average annual consumption. According to a May 2017 statement by Iran's Deputy Energy Minister, Sattar Mahmoudi, peak electricity consumption is expected to grow 7% in summer 2017 over the previous year.[467] Maximum power demand occurs in the summer, typically in July and August. The maximum demand in

---

464. BMI Research, *Iran Power Report*, Q4 2015, pp. 13-14 (note: the author has a copy of this report, which is available on line only at extremely high cost).
465. Hamed Shakouri and Allyeh Kazemi, "Electrical Energy Demand Forecast of Iran," https://www.researchgate.net/publication/280573416_Electrical_Energy_Deman d.
466. "Iran approves the Sixth Development Plan to Boost Investment," http://www.lexology.com/library/detail.aspx?g=9a25e9cc-44d0-45fa-9a54-905de434fc66
439. "Prediction of Increase in Peak Electricity Consumption for Summer," *Donya-e Eqtesad*, May 3, 2017, http://donya-e-eqtesad.com/SiteKhan/1203157

2015 was 53,654 MW, which MOE calculates as a 65.2% generation load factor.[468] Not all of that was met; the maximum supplied power was 49,351 MW. That said, Iran has much less of a problem than many other Middle East countries about how to meet peak load. Whereas in Saudi Arabia, for instance, peak summer load is vastly higher than winter load, in Iran a fair amount of power is used for winter heating (most Iranians live in the northern cities, which are quite cold in winter), so the winter load is a respectable fraction of the summer load.

Atomic Energy Organization of Iran (AEOI) President Dr. Ali Akbar Salehi has noted that the Majlis has charged the government to have 20,000 MW of nuclear power in a long-term horizon, and 8,000 in the medium-term, defined as 15 years.[469] In 2015, Salehi said that within the next 10-15 years, the number of nuclear power plants should reach eight and make up about 8-10% of electricity production. Salehi claimed that the cost per kWh of electricity generation from nuclear power was cheapest among all sources: nuclear – 10 U.S. cents; combined cycle – 12 cents; steam – 14 cents; and single-cycle gas – 19 cents. However, the validity of his claim is questionable. Salehi frequently refers to older studies, including some from the 1970s, which enthusiastically endorsed nuclear power, especially since Iran's oil fields had already markedly aged by the 1970s (since then, technological advances have done wonders to sustain production from oil fields, which had been expected to have been exhausted long ago).

More immediately, Russia and Iran signed a nuclear power cooperation agreement in November 2014 which was followed by a

---

468. MOE/Tavanir, *Statistical Report on 49 Years*, p 16.
469. "Production of 20,000 MW of Nuclear Power in the Long-Term," Hamvatan, September 11, 2016, http://www.hamvatansalam.com/news210307.html; "Plan to Produce 20,000 MW of Atomic Electricity," Donya-e Eqtesad, April 30, 2016 http://donya-e-eqtesad.com/SiteKhan/824089

September 2016 agreement, signed in a grand ceremony in Moscow, for the construction of two additional nuclear power plants at Bushehr, which also envisaged a further two, and four more at locations yet to be determined.[470] The plants at Bushehr would be imports from Russia; the others would be Iranian-built with Russian assistance. The agreement specified that all eight would use Russian fuel. Continuing the leisurely pace of implementing the 2014 agreement, a ceremony was held in Bushehr to mark the formal start of construction on March 14, 2017[471] – a week before the Iranian New Year, at a time when Iranian large-scale projects are frequently proclaimed to have begun when actual work does not get going for many months or even years later. At that ceremony, it was announced that the first 1,000 MW WWER reaction would be completed in October 2024 and the second in April 2026. It is worth noting that construction of the first Bushehr reactor took more than three times as long as the schedule announced when the project was launched.

Even if the ambitious announced schedule for completion of two additional nuclear power plants is met, their additional 2,000 MW capacity would be a small part of the additional capacity Iran is likely to build in the next decade. The Stanford study forecasts that additional capacity as 15,000 MW; the BMI study forecasts capacity as 30,000 MW; and the Iranian government's plan is for even more.

---

470. Andrew Kramer, "Russia Reaches Deal with Iran to Construct Nuclear Plants," *New York Times*, November 11, 2014, https://www.nytimes.com/2014/11/12/world/europe/russia-to-build -2-nuclear-plants-in-Iran-and-possibly-6-more.html?r=0; "Iran,Russia to Start Building Two Nuclear Power Plants," Radio Free Europe/Radio Liberty, September 2, 2016; "Iran to Produce 2,000 MW of new nuclear power on the Persian Gulf," Tasnim News, September 10, 2016, https://www.tasnimnews.com/fa/news/1395/06/20/1182687/-2-خیز-ایران-برای-تولید
هزار-مگاوات-برق-هسته-ای-جدید-در-حاشیه-خلیج-فارس.

471. "Construction begins on Iran's Bushehr 2 Nuclear Power Plant: Rosatom," *Tehran Times*, March 15, 2017, www.tehrantimes.com/news/412015/Construction-begins-on-Iran-s-Bushehr-w-Nuclear-Power-Plant

An additional 2,000 MW of nuclear power would be 13.4% of the lower estimate and 6.7% of the higher estimate.

Ineffective Past Reform Efforts

Any discussion of Iran's electricity sector should include a frank evaluation of the loudly proclaimed reform efforts over the last decade. While much heralded at the time – including by the International Monetary Fund (IMF) and World Bank – as models of how to move from wasteful to efficient policies, the reforms can in retrospect be called at best modest and more accurately failures.

For years, debate raged in Iran about the vast sums being devoted to subsidizing energy, including electricity (gasoline subsidies drew most of the attention). To the surprise of many, populist President Mahmoud Ahmadinejad embraced the far-reaching reforms recommended by international economists, namely, raising energy prices to world market levels and replacing energy subsidies with cash grants to families and firms to cushion the impact of the higher prices. A version of the reform passed the Majlis on January 5, 2010, which covered the full range of energy subsidies, including gasoline and natural gas as well as electricity.[472] The plan, developed in large part in consultation with the World Bank, was much heralded by the International Monetary Fund. In a September 2010 press release, the IMF noted that the suspension of oil and gas subsidies "should remove distortions and restore efficiency in the economy."[473]

---

472. Putting an End to the Subsidy Trap, *Financial Tribune*, January 20, 2015, https://financialtribune.com/articles/economy-domestic-economy/9392/putting-an-end-to-the-subsidy-trap

473. "IMF Survey: Iran to Cut Oil Subsidies in Energy Reform," International Monetary Fund, September 28, 2010. "Case Studies on Energy Subsidy Reform: Lessons and Implications," International Monetary Fund, January 28, 2013, http://www.imf.org/external/np/pp/eng/2013/012813a.pdf

However, the actual results differed considerably from the design intentions. For one thing, many Iranians live in apartment buildings, and it was common for the building to have only one meter – with the result that individual unit-dwellers had little incentive to be more efficient in using electricity. Even if that problem were resolved – and anecdotal evidence suggests progress on that front – the dramatic change from ending extensive subsidies hit hard at "legacy" consumers, that is, those consumers who made decisions on the basis of the past policies. An industrial firm whose cost structure is based on cheap electricity could well become unprofitable if forced to suddenly pay an unsubsidized electricity rate. For that reason, the original plan was to devote 30% of the higher revenue from the new rates to subsidized loans or grants to firms to invest in energy-savings.[474] But little, if any, of that money materialized. What funds were generated by the reform were insufficient to cover the generous grants to families that Ahmadinejad grandly announced – especially since he abandoned the overly ambitious plan to determine each family's income and then tailor the subsidies to the amount of income (a plan remarkably at odds with Iran's circumstances, given that few families file income tax returns and little is done to enforce the income tax laws). His political opponents suggested that Ahmadinejad's interest in the energy reform was motivated by his desire to claim credit for sending each Iranian family a large check each month.

---

474. "Initially, the benefits were to be paid in cash, while in a second phase, some of the additional revenues were to be used to support higher social benefits and public goods. Thirty percent of the additional revenues were to be used to assist Iranian companies restructure to adjust to the new, dramatically higher energy costs. The remaining twenty percent of the additional revenues would go to the government to cover government's own higher energy bill. Article 15 of the Reform Act authorized the government to establish a new Subsidy Targeting Organization to ensure efficient centralized management of the reform." https://www.imf.org/en/Publications/WP/Issues/2016/12/31/Iran-The-Chronicles-of-the-Subsidy-Reform-25044

The result of charging higher rates without compensating firms was that many industries would have gone bankrupt if forced to pay the new energy rates. It would appear that some firms simply did not pay – a strategy adopted by many consumers as well. The government has not published systematic data on unpaid electricity bills, but anecdotal accounts suggest that faced with soaring bills, consumers reacted with outrage and did not pay.[475] Given the many press reports about higher bills and given Iran's vigorous press, had there been numerous disconnections for non-payment, it would certainly have been reported and would have been deplored by populist politicians. Since no such thing occurred, it seems more likely that there were few disconnections and instead many arrears.

In any event, the new higher electricity rates were largely reversed by the collapse of the Iranian rial. Under the pressure of crippling international sanctions – the European cut-off in fuel purchases and U.S. measures which effectively prevented banks anywhere from handling payments for Iranian oil – the rial collapsed, first on the semi-legal open market and then at the official exchange rate. The official rate went from 10,616 rials to the dollar in 2011 to 18,414

---

475. According to *Tabnak* of August 15, 2010
http://www.tabnak.ir/fa/news/114238/%D9%82%D8%A8%D8%B6-%D8%A8%D8%B1%D9%82-%D8%B3%D8%A7%D8%B1%D9%88%DB%8C%D9%87%D8%A7-%D8%B1%D8%A7-%D9%82%D8%A8%D8%B6-%D8%B1%D9%88%D8%AD-%DA%A9%D8%B1%D8%AF, electricity consumers in the city of Sari, Iran protested the high electricity bills they received in the wake of the reforms. One consumer claimed that the price of power consumption was estimated to increase by 50-60 percent despite the fact that the Deputy Energy Minister said that electricity tariffs would be capped at a 10 percent increase. The same consumer claimed that his electricity tariff had increased 25 percent from the tariff he had paid previously. In less than two months, his electricity tariff had risen from 150,000 rials to 2,090,000 rials. Another resident of Sari noted that that number was wrong, however, claiming that his electricity bill had only risen from 180,000 rials to 1,610,000 rials. A third resident described a 40-50 percent increase in the price of electricity bills.

rials in 2013.[476] At the new value of the rial, the post-reform rates for gas and oil were suddenly well below international market rates, meaning that electricity generated with these fuels was once again heavily subsidized.

Added to which, the Ahmadinejad government reacted to the sanctions-induced shortfall in government revenue by resorting to the printing press. As the money supply soared, so did inflation. Consumer prices, by the official index, rose approximately 17% in 2014 alone and a cumulative 157% from 2011/12 through 2015/16.[477] The electricity rates were barely adjusted. From 2010-2011, they rose from 160 to 450 rials;[478] since then, they have barely budged, with the average 2016 rate being about 500 rials. The impact has been that electricity is once again heavily subsidized as it was before the reform. The International Energy Agency estimated that Iran's 2015 electricity subsidies were among the highest in the world, matched only by those in Russia and Saudi Arabia (see Chart 1.4, Energy Subsidies by Country 2015). Note that the same estimate shows Iran as providing such large subsidies for other types of energy that its overall energy subsidies were the largest in the world.

An additional evidence the reform was not as far-reaching as initially envisaged is that demand for electricity was little affected by the reform, despite the dramatic increase in announced rates. Gross power generation did not fall any year in the last decade; in the year of the maximum reform (2011), it rose 3.0%, little different from the growth in other years of the decade. In the period 2010-

476. Official exchange rate (LCU per US$, period average), The World Bank, http://data.worldbank.org/indicator/PA.NUS.FCRF?locations=IR
477. Consumer price index (2010 = 100), The World Bank, http://data.worldbank.org/indicator/FP.CPI.TOTL?locations=IR
478. "Energy Price Reform and Energy Efficiency in Iran," Saeed Moshiri, International Association for Energy Economics, 2013, https://www.iaee.org/en/publications/newsletterdl.aspx?id=197

2015, the average increase in gross power generation was 3.8%.[479] One could argue that a price reform would take a while to show up in consumption data, as consumers have few alternatives the day after the new rates take effect. But in a relatively short time – certainly within a couple years – consumers could shift habits and invest in energy saving technologies. The failure of demand to show much such effect certainly reinforces the view that the new rates were not effectively collected.

This dramatic and unsuccessful reform effort has left Iranian politicians with little enthusiasm for energy reform – and has made Iranian consumers highly skeptical that any announced reforms will in fact be implemented. Both politicians and consumers seem largely resigned to a continuation of current policies with only minor modifications implemented slowly. That said, it is worth pointing out the extraordinary opportunities for Iran to save money through a variety of non-nuclear avenues. Cutting subsidies is the most obvious but most difficult measure. Other steps involve changing the investment decisions, such as relying more on renewable energy and moving more towards combined cycle generation rather than single cycle. And even if those steps are not taken, it still remains the case that Iran would do better to generate in natural gas-fired facilities rather than in nuclear power.

Cutting Subsidies

The failure of earlier reform efforts has left Iran with heavily subsidized electricity rates. The most obvious way to meet Iran's future electricity needs is to reduce those subsidies and therefore

479. *Statistical Report on 49 Years*, p. 10. Net power generation (p. 12) followed the same pattern.

reduce consumption. This is entirely unlikely, especially given the experience of the earlier dramatic reform effort.

One element in the subsidized electricity rate is the subsidized price of fuel. Consider natural gas. In theory, electricity producers pay about the same rate for natural gas as do other consumers, as explained by the Stanford Iran 2040 paper:[480]

> "In rough terms, natural gas in Iran is uniformly priced at $34,000/mcm for residential, power, and industrial consumers (including petrochemical plant fuel) despite the profound differences in the price sensitivities of these consuming sectors. For the sake of comparison, the above price is almost 50% lower than the Henry Hub gas spot price ($71,000 in 2016) and 90% lower than the average residential gas price in the United States ($355,000/mcm in 2016)."

While there is only an imperfect world market in natural gas, the price paid for natural gas by different consumers around the world has been converging. The differences were as much as five-to-one a few years ago what with the U.S. gas glut driving down prices in mid-America and with Japanese firms locked into long-term contracts.[481] Now, prices differentials are generally less than two-to-one. Even at the low end of that price range, Iranian electricity companies are paying less than half the market value of gas.

But that is not the end of the subsidies for electricity. Even if the fuel used were free, the price of electricity in Iran to consumers does not cover the cost of generation and distribution. Iranian press reports, citing government officials, refer to the cost of generation and distribution – without any allowance for the fuel – as 1,000 rials per kilowatt-hour while the average price paid by consumers is 500

---

480. Azadi, et al., "The Outlook for Natural Gas, Electricity and Renewable Energy in Iran," p. 11.
481. BP, *Statistical Review of World Energy 2017*, p. 33.

rials.[482] The end result is that the average price for electricity in Iran, at 500 rials per kilowatt-hour, is one-eighth the U.S. average: 1.5 cents compared to about 12 cents in the United States.

One would think that the privatization of much of the electricity generation capacity would create a powerful lobby group for more realistic electricity rates. After all, it is hard to see how the private generators make a profit if the average rate is half the cost of generation and distribution even were fuel free. However, Iranian "privatizations" have a peculiar character. For one thing, they are remarkably opaque. In June 2017, President Rouhani said that no more than 13% of the assets privatized went to truly private hands.[483] Many "privatized" assets ended up in the hands of those connected to the Islamic Revolutionary Guard Corps (IRGC). In the same speech, Rouhani characterized the privatization to date as transferring assets from "the government without guns" to "the government with guns" – a much quoted comment by Iranian commentators. In addition, other assets went to quasi-government entities, such as the many retirement funds run by ministries and state-owned enterprises, or the funds in each province which are supposed to transfer a large chunk of privatized assets to the poor. Since if those funds actually handed the assets to the poor, the poor would promptly sell them, driving down the price, what has happened in practice is that the central government manages those funds theoretically on behalf of the poor. It would appear that much of the privatization process has been run in a way which allows the politically well-connected to divert government funds into their pockets.

---

482. "How Much is the Cost of Production of Electricity?" (in Persian; translation by author), *Donya-e Eqtesad* (the Iranian financial newspaper), October 14, 2016, http://donya-e-eqtesad.com/SiteKhan/1129248.
483. "President of the Republic: Everyone Should Trust the Private Sector," (in Persian – author's translation), June 24, 2017, http://www.etelaat.com/etiran/?p=290910&print=0.

Certainly, the non-transparent way in which privatization has worked has been an impediment to attracting the foreign investment Iran claims it wants for infrastructure projects. Added to which, the active role of IRGC-connected figures creates a problem for any foreign investor who does any transactions in U.S. dollars, since the U.S. government retains strict sanctions – including large penalties – aimed at foreign firms that use U.S. dollars in transactions that are linked to IRGC-related firms. These factors will be a significant impediment to attracting foreign or domestic investment for expanding electricity generating capacity. Arguably, the barriers could slow non-nuclear projects – gas or renewable – so much that they could take about as long to build as do nuclear power plants, which are fully government financed.

Relying on More Renewable Energy

The main form of renewable electricity generation in Iran is hydropower. The Shah added much to Iran's hydro operational capacity, which went from 309 MW in 1967 – 36% of the total operational capacity for all sources – to 1,804 MW in 1978, 28% of total capacity.[484] After making only minor additions to hydro capacity in the first two post-revolutionary decades – hydro capacity going from 1,804 MW at the time of the 1978 revolution to 1,999 in 1998 –Iran has built quite a number of hydroelectric dams in the last two decades. From 2001 to 2015, installed operational hydro capacity rose sharply, from 1,999 MW to 11,278 MW, which meant that hydro went from 8% of total capacity to 17%.

At the same time, the government has encouraged increased irrigation with wasteful technology, with the inevitable result that water shortages have become quite common. Protests about water

---

484. All data in this paragraph are from MOE/Tavanir, *Statistical Report on 49 Years*, p 8.

shortages have broken out repeatedly in various parts of the country, including protests against the building of new hydroelectric dams. It seems very unlikely that Iran would be able to continue increasing the share of hydro in electricity generation; quite the contrary, that share seems slated to decline.

In 2016, a private estimate of the non-hydropower renewable energy sector in Iran – not fully compatible with the MOE/Tavanir data cited elsewhere – was comprised of mainly wind (53.88 MW) and biomass (13.56 MW), with much smaller amounts of solar (0.51 MW).[485] In 2016, the head of the Ministry of Energy (the 'MoE') stated that Iran was planning to tender 1,000 MW of wind and as much as 3,000 MW of solar projects.[486] MOE will purchase all electricity generated from renewable sources by approved private sector projects at specific feed-in tariffs (FiTs). The Sixth Five-Year Development Plan calls for 5,000 MW of renewable generating capacity. Tavanir has forecast that renewables will provide 10% of Iranian power by 2021. [487] Both of these forecasts seem implausibly optimistic.

To be sure, Iran has excellent natural conditions for solar power in particular. Iran is an arid country with clear skies much of the year and with large desert areas for which there are few economic uses

---

485. CMS Cameron McKenna LLP, *2016 Report on Renewable Energy in Iran* http://www.satba.gov.ir/suna_content/media/image/2017/02/5196_orig.pdf?t=636219021775330000. The report is sloppy on some matters, e.g., it mischaracterizes the data it presents on hydropower, which on close examination turn out to be data only on micro-hydropower projects.
486. CMS Cameron McKenna LLP, *2016 Report on Renewable Energy in Iran* http://www.satba.gov.ir/suna_content/media/image/2017/02/5196_orig.pdf?t=636219021775330000
487. Renewable Energy Trade Mission Iran http://www.retrademissioniran.com/news-articles/2016/9/15/who-will-fund-irans-first-batch-of-renewable-energy-projects. Among the foreign companies with which Iran had by September 2016 signed agreements to support renewable energy development were British Photovoltaic Association, Genesis of Italy, and Shanxi Energy of China. Renewable energy projects in Iran: Solar Focus, October 21, 2016, Dentons.

other than solar power. Given this paper's focus on Iranian institutional arrangements and economic factors, this is not a subject which will be explored in more detail here.

Updating the Technology Used to Generate Electricity with Thermal Fuels

Iran makes extensive use of steam generation rather than gas turbines. Nearly everywhere else in the world, steam is seen as an inferior technology to gas turbines. Iranian sources are more enthusiastic about steam. Indeed, one Iranian study ranked steam as the cheapest source for electricity generation.[488] It is not clear why steam is rated highly in Iran. Perhaps the answer is related to the need to use fuel oil during the winter months in so-called "gas" facilities. Iran's fuel oil is high in sulfur. Burning high-sulfur fuel oil is presumably not good for a turbine. Also, that fuel oil has few other economic uses and so is presumably cheap. Fuel oil is not in high demand internationally; indeed, what market does exist for the fuel oil will largely disappear as the global shipping industry makes the shift (required by nearly all countries as part of an international agreement) to phase out use of this type of fuel in ships by 2020. Anyway, if Iran were able to work out arrangements – perhaps including guaranteed year-round gas supply – to make steam less attractive and were Iran to phase out its large stock of aging steam

488. Khakbazan-Fard et al., "Generation Expansion Planning for the Iran Power Grid." Three of the studies four authors work for Tavanir. They say their analysis is based on the WASP IV software package, which they characterize as "one of the well-known power expansion planning softwares." They conclude, "For supplying the base load, steam power plants are chosen. The main reason is the lower price of the fuel for steam plants (9 months gas and 3 months HFO [heavy fuel oil]) in comparison to the fuel of gas and combined cycle power plants (9 months gas and 3 months gasoil)."

facilities, it would seem Iran could gain considerable efficiencies in electricity generation.

Another issue is single-cycle versus combined-cycle gas generators. In a single-cycle gas generator, air is filtered in to an air compressor. Natural gas is introduced to a burner, where in the presence of the compressed air, it produces a high temperature and high pressure gas that spins a turbine connected to a generator, producing electricity. In a combined-cycle generator, extra electricity is produced by introducing water, where exhaust gas from the original output is used to heat water into a steam, where it turns a second turbine, generating more power through the generator, after which it is condensed and sent back to the exhaust gas chimney. A combined-cycle generator will produce up to 50% more power than a single-cycle generator.[489]

Since 2005, single-cycle gas turbine generation has grown 57.4%, while combined cycle has grown 64.1%. In 2015, combined cycle generated 25.3% more power than single cycle gas turbine generation. There may well be scope for more efficiencies from shifting to more combined-cycle facilities, as Iran is doing[490]. However, it should be noted that single-cycle facilities are better suited for firing up quickly to meet peak demand, so there will also be room for some such facilities in the best-managed system.

---

489. "Iran Increases Combined Cycle Power Output," *Financial Tribune*, July 13, 2016, https://financialtribune.com/articles/energy/45349/iran-increases-combined-cycle-power-output

490. "Iran Increases Combined Cycle Power Output," *Financial Tribune*, July 13, 2016, https://financialtribune.com/articles/energy/45349/iran-increases-combined-cycle-power-output; "Iran Starts Construction of 890 MW Combined-Cycle Power Plant," *Financial Tribune*, January 28, 2017, https://financialtribune.com/articles/energy/58312/iran-starts-construction-of-890-mw-combined-cycle-power-plant; "Two Combined-Cycle Power Plants Launched in Southern Iran," *Financial Tribune*, February 8, 2017, https://financialtribune.com/articles/energy/59113/two-combined-cycle-power-plants-launched-in-southern-iran\

Relying on Natural Gas

Despite some complications, Iran has much potential to continue relying on natural gas as the principal fuel for electricity production, which is precisely what all serious studies recommend and what the government has announced. Even assuming that the government's optimistic schedule for new nuclear power plants and renewable energy comes to pass, Iran will still rely on natural gas for most of the additions to its electrical generating capacity.

In recent decades, Iran has dramatically increased its reliance on natural gas. In 1978/79, Iran's electricity generation capacity was 7,086 megawatts (MW), of which natural-gas-fired facilities were only 2,523 MW. As noted before, in 2015/16, the total capacity was 64,708 MW, of which natural gas fueled 21,630 MW. In other words, of the additional 57, 622 MW in capacity, 19,107 MW was fueled by natural gas.

That fits with Iran's general pattern of making intensive use of its natural gas. In 1979, Iran's overall gas consumption – the vast majority being outside of electricity generation -- was 3.6 million barrels of oil equivalent (MBOE) compared to 83.8 MBOE of oil consumption. In 2016, gas was 180.7 MBOE and oil was 83.8 MBOE. In other words, Iran shifted from being a country which used little natural gas to one which uses a great deal.

Few countries in the world rely as heavily as Iran on natural gas as a proportion of their primary energy consumption. In the BP data, the share of natural gas in primary energy in Iran is 66.7 percent; in the United States, 31.5 percent; in the European Union, 23.5 percent; and in Russia, 52.2 percent. [491] The only countries in the same league as Iran for relying heavily on natural gas are Turkmenistan, Uzbekistan, Qatar, and Trinidad and Tobago. Natural gas in Iran is

---

491. BP, *Statistical Review of World Energy 2017*

used extensively, not only in heating and industrial uses, but even in ground transportation; the Iranian vehicle fleet has the world's highest share of vehicles power by compressed natural gas.[492] About 30% of Iranian gas consumption is for electricity generation; 30% is for space heating; 30% is for industrial uses, especially petrochemicals; and 10% is for reinjection into aging oil fields to boost pressure.[493]

A major reason that Iran makes such extensive use of natural gas is that it has extensive reserves. According to BP, Iran's natural gas reserves in 2017 are 33 trillion cubic meters (TCM), making up 18.2% of the global reserves.[494] However, a word of caution is in order about reserve data. There is no agreed international standard for what constitutes reserves, and Iran, like many countries, is less than transparent about it calculates the reserves figures it announces (BP largely relies on figures announced by governments). The U.S. reserve figures are much more cautious than those of most countries, being calculated by strict standards designed to let investors know what companies' prospects are. Furthermore, it is not apparent why a commercial company would invest in finding reserves which are not likely to be used for many years, so it may well be the case that U.S. firms satisfy themselves with a cushion of several years' reserves and do not invest much in looking for additional reserves. In other words, figures on reserves can be a poor indicator of a country's production potential.

In many ways, what is more relevant in judging Iran's prospects for increasing gas production is its historical record. As noted earlier,

---

492.Indeed, according to *Automotive Megatrends*, https://www.automotiveworld.com/analysis/back-business-iran-new-emerging market/ Iran has the second largest fleet of natural gas vehicles in the world.
493. Azadi et al., "The Outlook for Natural Gas, Electricity, and Renewable Energy in Iran."
494. BP, *Statistical Review of World Energy 2017*. The data on 1978 consumption compared to 2016 come from the underpinning tables published along with that report.

Iran has dramatically increased its production of natural gas from 3.6 MTOE in 1978 to 180.7 MTOE in 2016. This expansion has been particularly marked in recent years; 2011 production was 143.9 MTOE, meaning that production rose in 2011-2016 on average 4.7% per annum. However, Iranian MOE data show an odd 200-2013 dip in use of natural gas as fuel for electricity generation. The data from which Chart 1.3 is said to be drawn show gas used for electricity generation at 43.4 billion cubic meters (BCM) in 2009 falling to 36.6 BCM in 2013, before then jumping upwards to 50.2 BCM in 2014 and 58.4 BCM in 2015. This is roughly consistent with the BP data on Iranian gas consumption in all sectors, which shows rather slower growth in 2009-2013 before jumping in 2014-2015. But it is by no means clear why this may have happened. Reasons could have included problems in gas production or excessive amounts of fuel oil and diesel, which Iran had problems marketing abroad.

The domestic market looks to be the best place to use the increasing Iranian production of gas, given that prospects for exporting gas look poor. Besides political sensitivities with neighbors Turkey and Pakistan, Iran has not offered attractive price terms and has had great difficulties securing financing for pipelines. As for exporting liquefied natural gas (LNG), Iran does not and is not likely to get access to U.S. technology for such key components as gas compressors. Even if that problem can be bypassed as alternative suppliers develop better LNG technologies, Iran faces the problem of an extraordinary increase in LNG capacity coming in the not-too-distant future from Australia and then, a few years later, from the United States. In a well-supplied market, Iran would have difficulty competing – both for political reasons and because Iran's political culture makes it difficult for any Iranian official to agree to terms attractive to foreign purchasers without facing charges at home that they sold out the nation's interests.

While Iran has had great success over the years at increasing its gas output, it must be acknowledged that Iran has not been able to meet peak demand in the winter – in part because Iran has done a very poor job at constructing storage facilities where gas could be stored in the summer and then drawn down in the winter (reports circulate that Iran's storage capacity is on the order of a week of consumption, compared to much more in other countries with similarly harsh winters). And Iran will face constraints expanding its gas production because of the high capital requirements and Iran's limited access to international financial markets. That said, Iran continues to flare large amounts of natural gas rather than gathering the gas for productive use. The Stanford study estimates the flared gas at 33,000 MCM/day.[495] That matches the estimate from the World-Bank-run Global Gas Flaring Reduction Partnership in conjunction with major energy companies, which uses a satellite to estimate flaring around the world; its estimate of Iran's flaring in 2015 was 12,000 million cubic meters for the year, or 34,000 MCM/day.[496] That is equal to about 6% of Iran's current gas production; alternatively, that is 20% of the 58,154 MCM of gas used by Iran's power plants in 2015.[497] The Stanford study places the cost of capturing now-flared natural gas at $60,000-80,000 per MCM. In other words, by ending flaring, Iran has low-cost means to increase its gas output sufficient to meet the additional needs of electricity generation by efficient combined-cycle facilities.

To be sure, one can make the argument that Iran might do better to use much of the gas it can produce in the next few decades to re-inject into its aging oil fields, thereby boosting oil output while preserving the gas to be used several decades from now. The

495. Azadi et al., "The Outlook for Natural Gas, Electricity, and Renewable Energy in Iran," pp 11-12.
496. Global Gas Flaring Reduction Partnership, www.worldbank.org/en/programs/gasflaringreduction#7.
497. Gas production from BP *Statistical Review*; fuel used for electricity generation in 2015 from *Statistical Report on 49 Years*, p. 14.

Stanford Iran 2040 study was only the latest study pointing out that such re-injection is one of the highest return uses of Iran's gas; it calculated the value per MCM from re-injection as $200,000 compared to $50,000 from electricity generation or space heating, which did not even include the value of the gas which can be recovered in future years after oil has been extracted from the field into which the gas is reinjected.[498] Re-injecting gas has been recommended by Iranian technocrats since the Shah's days, but Iran has for forty years failed to act meaningfully to take advantage of this obvious way to sustain and increase oil production. Iran has spent many tens of billions of dollars on new investment in oil production which could have been saved had it instead acted to reinject natural gas in the aging fields. The failure to end flaring has been arguably as, if not more, costly than all of the policy decisions Iran has taken about electricity production, including the investment in nuclear energy.

Environmental Considerations

Iran's major cities, especially Tehran, suffer from dreadful air pollution. The vast majority of the pollution comes from the use of oil, especially gasoline and diesel in vehicles. Iran's vehicle fleet has rudimentary pollution control technology, and Iran's gasoline is of a low standard. Because relatively little of Iran's air pollution comes from electricity generation, any measures to reduce electricity demand or produce electricity from less polluting sources than the natural gas now used to produce most of Iran's electricity would

---

498. Azadi et al., "The Outlook for Natural Gas, Electricity, and Renewable Energy in Iran," pp 12-13. Admittedly, the value of CNG for ground transport, at $450,000 per MCM, and of exported gas, at $270,000 per MCM, were higher than reinjection, but the study notes the constraints on expanding use of gas for either of those purposes. The value of gas for industrial purposes ranged from $50,000 to $155,000 per MCM, with the higher figure being for petrochemical feedstock.

have little impact on Iran's urban air pollution problems. Specifically, any benefit from using nuclear power to generate electricity would have modest to negligible impact on Iran's air quality problems.

In many countries, the waste produced by nuclear power has become an environmental headache. That is not and probably would not be the case in Iran. Iran has been and likely will for the near future be dependent on nuclear power plants built by Russia. Under considerable international pressure – not least from the United States – exerted because of concern that reactor waste could provide Iran the basis for reprocessing plutonium for nuclear weapons, Russia requires Iran to surrender in a timely manner all nuclear waste materials to Russia. One could argue that Iran's use of nuclear power contributes to the global problem of nuclear waste disposal, but it is not apparent that Iranian authorities would factor into their decision-making the impact that they may have on a global problem as distinct from an Iran-specific problem.

During the negotiations of the JCPOA, Iranian officials repeatedly stated that Iran intended to design its own power plants which would use the enriched uranium fuel Iran was producing.[499] That was in response to repeated Russian statements that Russia would only sell power plants to Iran if all the fuel came from, and was returned to, Russia – a pledge meant to reassure Western countries there was little proliferation risk from Russian power plants sold to Iran. However, Iranian officials have said that in the negotiations about two additional power plants, Russia has been willing to consider Iran providing the fuel. This has not been confirmed by the Russian side. But it is possible Iran will provide some supplier – perhaps China – who would be willing to provide Iran a power plant that is to be fueled by Iranian-produced enriched uranium. If so, then Iran

499. "Annex I – Nuclear-related Measures," US Department of State, p. 2, https://www.state.gov/documents/organization/245318.pdf

would have to consider how to dispose of the fuel. There does not seem to have been any discussion in Iran about this issue. In this context, it is worth noting that Iran has two vast deserts of salt flats - perhaps a fuel disposal site could be located in one of these.

Why the Enthusiasm for Nuclear Power?

Obviously one reason Iranian officials are enthusiastic about nuclear power is the potential it gives for a nuclear weapons option. However, additional factors are also at play. Those include attraction for what is perceived as advanced technology and also the institutional arrangements in electricity generation.

Attraction of Advanced Technology

One of the main themes in Iranian Supreme Leader Khamenei's public statements is the importance of Iran's scientific advance and what he sees as Western opposition to Iranian scientific progress:

> "Colonial countries are quite aware of the fact that in order to keep a country under their political and economic dominance, they should bar its scientific progress."[500]

> "They are opposed to the progress and development of the Iranian nation. They do not want an Islamic and independent country to achieve scientific progress and possess advanced technology in the Middle East region, a region which possesses most of the world's oil and which is one of the most sensitive regions in the world. They are worried about anything that can help the regional nations to achieve independence, self-reliance

500. Address to university professors and elite academics, October 13, 2005. (http://carnegieendowment.org/files/sadjadpour_iran_final2.pdf)

> and self-sufficiency. They want this populous region, which is rich in mineral resources, to be in need of them forever. This is why they are opposed to our possessing modern technology and to our youngsters making progress in scientific areas. It is hard for the global arrogance to accept that the talented Iranian nation has been able to take great strides in the field of science and technology, especially in the field of nuclear technology. They want Iran's energy to be always dependent on oil, since oil is vulnerable to the policies of world powers. They aim to control other nations with invisible ropes."[501]

Khamenei sees nuclear power in this light. That is more than a bit peculiar, given that nuclear power is a 60-year-old technology which has not proven particularly successful in the most scientifically advanced countries. But Khamenei is not the most up-to-date in his praise of technology and science; for instance, he rarely refers to software and computer technology. Besides scientific progress, Khamenei's other fixation is on self-reliance. Khamenei has made numerous statements over the past few years highlighting Iran's need for economic self-reliance, which, in his eyes, will enable Iran to be politically independent. *"Iran should not be pinning its hopes on when the enemy will lift the sanctions...the hell with them. We should look into what we can do ourselves."*[502] One might suggest this is a bit odd, given that so much of Iran's economic well-being comes from oil exports. And one might ask how reliance on nuclear power promotes economic self-reliance, given that Iran has no capacity to make the fuel rods needed for a nuclear power station and is not likely to be able to do so for a long time to come. But in

---

501. Supreme Leader's address to high school students, March 14, 2005. (on the United States) (http://carnegieendowment.org/files/sadjadpour_iran_final2.pdf)
502. "Iran's leader calls for self-reliance in face of sanctions," 2014, Middle East Institute, mhttp://www.mei.edu/content/irans-leader-calls-self-reliance-face-sanctions

Khamenei's mind, nuclear power is connected to economic self-reliance.

Institutional Arrangements Favoring Nuclear Power

Certain institutional arrangements made the original decision to pursue nuclear power more attractive, and other institutional factors proceeded to reinforce the original choice.

To start, there was the initial decision to invest in nuclear power. One factor that may have been involved was that the institutional separation between the electricity generator – i.e., at the time, entirely MOE – and the fuel supplier. MOE does not control the supply of gas for the electricity plants. As explained elsewhere, gas supplies are interrupted in the winter when demand is at its peak. Such interruptions are particularly likely to hit large consumers, such as electricity generating stations, rather than the household network; as noted elsewhere, electricity generation takes about 30% of gas used in Iran.

Whereas the electricity generators may not be given priority in gas distribution, by contrast, the Bushehr nuclear power plant does not have to worry about fuel interruptions. Bushehr, which is owned and operated by the Iranian government through the Atomic Energy Organization of Iran, has a guaranteed fuel supply. And Bushehr does not have to worry about being cost-competitive with other electricity generators, since the decisions about its operations do not seem to be much influenced by cost. [503]

An additional institutional factor which promotes nuclear energy is the momentum created once nuclear power starts. Once authorities

503. "Iranian-Run Bushehr Plant has IAEA Oversight," Voice of America, November 2, 2009, https://www.voanews.com/a/a-13-2009-06-10-voa60-68826397/413931.html

made the investment to create a large atomic energy agency, it became a powerful lobby for its self-perpetuation and aggrandizement. In other words, the decision to pursue robust nuclear facilities launched Iran on a path from which departure becomes harder and harder. The organized pressure to sustain nuclear energy comes in no small part from the engineering lobby. Consider that the unemployment for engineering graduates in Iran is, at 22%, higher than that that for holders of degrees in the humanities and education, which was 15% or less.[504] It is not clear what the unemployment rate is for nuclear engineers.

And then there is natural propensity of politicians to validate the decisions they made in the past. Politicians are no more willing than the rest of humanity to acknowledge that they were mistaken. Having spent on nuclear power billions of dollars – and having suffered years of crippling international sanctions which Iran explained (largely incorrectly) as being designed to impede Iran's nuclear power program – the politicians who made those decisions would have great difficulty bringing themselves to acknowledge that the pursuit of nuclear power was a mistake. The nuclear power plans were supported by every major Iranian politician, including the vigorous support of Supreme Leader Ali Khamenei, who has been in that post since 1989 and seems set to hold it until he dies.

Overall Evaluation

This study has shown that Iran has many ways to meet its electricity needs which are economically more attractive than use of nuclear energy. That said, Iran's electricity system is generally badly run from an economic perspective. While nuclear power is a poor use of

504. Nader Habibi, "Iran's Overeducation Crisis: Causes and Ramifications," Crown Center for Middle East Studies, Brandeis University, February 2015, https://www.brandeis.edu/crown/publications/meb/MEB89.pdf (P.4)

resources for Iran, the construction of additional nuclear power plants would not make the list of the most wasteful things Iran has done regarding electricity. Furthermore, the powerful political pressures to continue with nuclear power – to validate past decisions, to provide employment to highly trained engineers, to fit with Khamenei's "resistance economy" rhetoric – make it extremely unlikely Iran will revisit the decision to proceed with additional power plants.

The Enigma of Resource-Rich Iran's Pursuit of Nuclear Energy

Geoffrey Styles

I. Introduction

When Iran agreed to the Joint Comprehensive Plan of Action (JCPOA) in July 2015, it cleared a path for the Islamic Republic to achieve an ambition that predates that government's existence,[505] at least for the non-military applications of nuclear energy. The terms of Iran's agreement with China, France, Germany, Russia, the United Kingdom, United States and European Union permit it to build civilian nuclear power plants while continuing to develop uranium enrichment under UN supervision.

However, the JCPOA left unaddressed the larger question of the appropriateness of nuclear power in a country with the world's second-largest reserves of natural gas, along with abundant, untapped renewable energy resources, all of which appear to be more economical in the Iranian context than the envisioned fleet of nuclear power plants.

Iran has already paid a high price for its lengthy pursuit of nuclear energy. That price has included both direct costs in the form of foregone oil exports under international sanctions, along with substantial indirect costs resulting from reduced or restricted access

505. https://www.cfr.org/backgrounder/international-sanctions-iran

to foreign direct investment, lending, and the settlement of dollar-denominated transactions.[506]

The Iranian nuclear program attracted international scrutiny from its inception, particularly in light of Iran's decision to pursue not just the deployment of nuclear power plants, but of the nuclear fuel cycle starting with enrichment. Over many years, intelligence gathered by the U.S. and other nations raised concerns that Iran's nuclear ambitions extended beyond civilian applications to include the development of a nuclear weapons capability. These suspicions were bolstered by periodic discoveries of undeclared nuclear facilities such as the Arak heavy water reactor[507] and Natanz uranium enrichment plant in 2002[508] and the underground enrichment facility at Fordow in 2009.[509]

The prospect of Iran becoming a nuclear weapons state alarmed the international community and prompted successive rounds of increasingly strict United Nations sanctions, beginning in 2006.[510] At their most restrictive point, the sanctions targeted Iran's oil industry and oil exports, the country's principle source of foreign exchange and the engine of its economy. Other sectors and designated groups were also affected, including entities closely linked with the economic interests of Iran's leadership.[511]

From late 2011 the combination of U.S. and UN sanctions had a severe effect on Iran. As seen in Figure 1 below, the Iranian economy contracted in three of the four years the strictest sanctions

---

506. https://www.jdsupra.com/legalnews/ofac-strengthens-iranian-embargo-by-revo-81420/
507. http://www.nti.org/learn/facilities/177/
508. http://www.economist.com/node/14529841
509. https://www.reuters.com/article/us-nuclear-iran-fordow-plant/factbox-irans-fordow-plant-idUSBRE84N0MB20120524
510. https://www.cfr.org/backgrounder/international-sanctions-iran
511. https://www.wsj.com/articles/hit-ayatollah-khamenei-in-his-pocketbook-1516666405

were in place. During this period the country significantly underperformed[512] the broader Middle East and North Africa region, which otherwise enjoyed steady, moderate growth.

Figure 1:

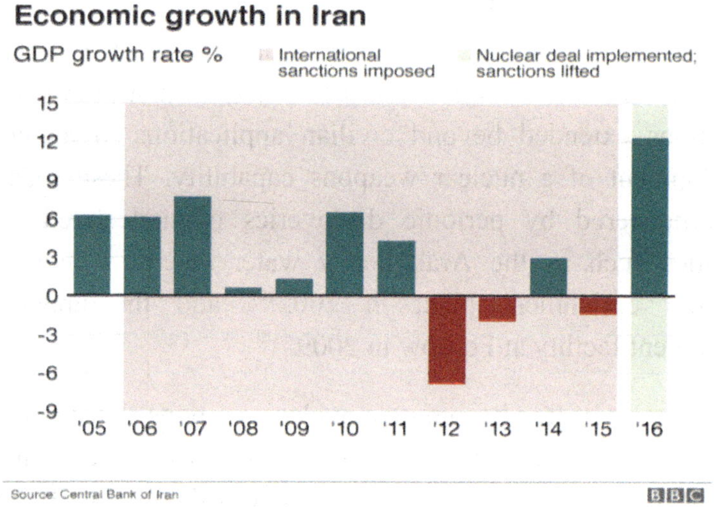

Considering only the impact on Iran's global sales of crude oil, its oil exports were cut by around 1.3 million barrels per day.[513] An accurate estimate of total foregone oil profits on these volumes is impractical, since some of this oil was produced and stored for later sale, while other production was shut in, some of which may have been lost permanently. However, as reported by OPEC,[514] the cumulative reduction in Iran's oil revenues in 2012-15, compared to 2011, came to $215 billion. Even after excluding the portion of that attributable to lower international oil prices in 2014 and 2015, the

512. https://www.ft.com/content/b6994c78-36f3-11e7-bce4-9023f8c0fd2e?desktop=true&segmentId=7c8f09b9-9b61-4fbb-9430-9208a9e233c8

513. https://fas.org/sgp/crs/mideast/RS20871.pdf

514.http://www.opec.org/opec_web/static_files_project/media/downloads/publications/ASB2016.pdf

cost to Iran's economy still exceeded $175 billion, equivalent to around 5 months of current GDP.

Nor were sanctions the only pressure exerted on Iran in conjunction with its nuclear program. From the mid-2000s until 2013's "Joint Plan of Action,"[515] nuclear negotiations with Iran ebbed and flowed without a clear prospect of reaching resolution. The determination of the Iranian government not to give up or sharply constrain the activities in question fueled speculated that Iran's nuclear ambitions could only be curbed through military strikes,[516] whether by the US, an international coalition, or regional powers. This prospect appeared sufficiently likely to influence global oil markets by means of an "Iran risk premium"[517] that helped to push oil prices above $100 per barrel on their way to an all-time high of $148 in 2008.

Iran's agreement in 2015 finally to limit its efforts, particularly with regard to enrichment, has been rewarded with the lifting of the nuclear-specific sanctions on the oil and gas industry along with many of those affecting other sectors. Some sanctions related to other aspects of Iran's policies and regional actions remain in place, but commercial ties, particularly with European countries and companies, have rebounded. Foreign Direct Investment increased to $12 billion[518] in 2016, or roughly five times the average of 2003-2010. More than two-thirds of that investment was in oil and gas exploration, production and processing.

---

515. https://www.armscontrol.org/Implementation-of-the-Joint-Plan-of-Action-At-A-Glance
516. http://www.nytimes.com/2006/01/22/weekinreview/why-not-a-strike-on-iran.html
517. https://blogs.wsj.com/marketbeat/2011/11/08/nymex-crude-gushing-toward-100-partly-because-of-iran/
518. https://www.ft.com/content/b6994c78-36f3-11e7-bce4-9023f8c0fd2e?desktop=true&segmentId=7c8f09b9-9b61-4fbb-9430-9208a9e233c8

Meanwhile, Iran's single existing light-water reactor is in operation at Bushehr, with technical assistance from Russia. The facility is essentially a Russian modification of the previously uncompleted, pre-Iranian Revolution design and construction by Siemens[519] of Germany, completed in 2011 by Russian contractors. Under the terms of Iran's nuclear power agreement with Russia, the latter is responsible for fueling this reactor and will receive all spent fuel from its operation. The second of as many as eight additional Russian reactors is reportedly under construction[520] at the same site.

Iran's history of extraordinary costs incurred in pursuit of nuclear energy, along with its ongoing dependence on foreign nuclear expertise and fuel supplies, raises serious questions about the suitability of nuclear energy for the county. The county's economy has been built around its prodigious endowment of hydrocarbon resources, which meets the bulk of Iran's domestic energy needs and provides most of its income from international trade. At this point, nuclear power and even renewable energy still represent merely asterisks in its energy balances. Does Iran really need nuclear power plants to fuel its economic growth, or does it have better, more economical and less controversial options? That is the fundamental question this paper seeks to answer.

II. Energy Demand

As indicated in the chart below (Figure 2) the energy mix of the Islamic Republic of Iran is dominated by two related fossil fuels: petroleum and natural gas. Together these meet 98% of the country's energy demand, supplemented by modest contributions

---

519. http://www.world-nuclear.org/information-library/country-profiles/countries-g-n/iran.aspx
520. https://www.reuters.com/article/us-iran-russia-nuclearpower/iran-russia-start-construction-of-new-iranian-nuclear-plant-idUSKCN11G0EB

from coal, hydropower and other renewables, and the output of country's only commercial nuclear power plant.

Figure 2:
Share of total primary energy supply* in 2015

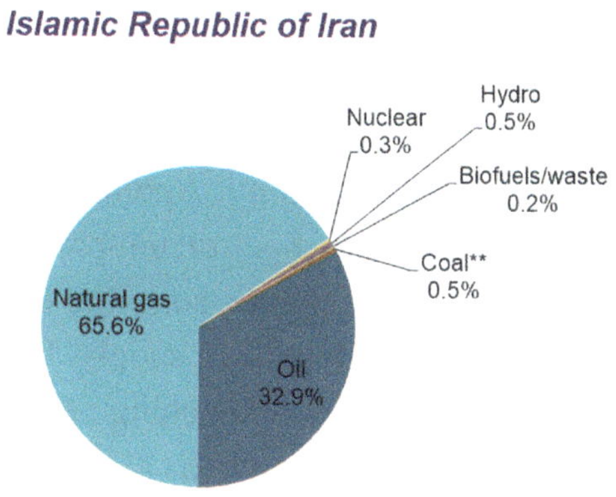

Source: International Energy Agency

As discussed in greater depth below, Iran has large reserves of both oil and natural gas. However, more than half of Iran's crude oil production is exported, while nearly all gas production is consumed internally. Gas is the mainstay of the domestic economy, with a 66% share of energy demand[521] in 2015. It is used directly in the commercial and industrial, residential and transportation sectors. Iran is unusual among oil exporters in having nearly a quarter[522] of its automobiles running on compressed natural gas (CNG). Gas also fueled the bulk of Iran's electricity supply in 2015.[523]

In considering the current and potential future role of nuclear power in Iran's energy market, it is essential to examine where and how it competes with other energy sources. The main output of commercial

---

521.http://www.iea.org/statistics/statisticssearch/report/?year=2015&country=IRAN&product=Balances
522. http://www.ra.camcom.gov.it/eurosportello/allegati-pina-newsletter/iran-automotive
523.http://www.iea.org/statistics/statisticssearch/report/?year=2015&country=IRAN&product=ElectricityandHeat

nuclear power is electricity. In this respect nuclear power does not compete with the consumption of fossil fuels by motor vehicles or for process heat in industry. Space heating, transportation and other applications are amenable to electrification, but not to direct fuel substitution by nuclear energy.

Iran consumed 221,000 Gigawatt-hours (GWh) of electricity in 2015. As shown in Figure 3 below, facilities burning natural gas supplied 79% of that electricity. Another 14% was generated from oil—mainly seasonal use of petroleum products such as high-sulfur fuel oil—with all other sources contributing just over 6%. Within that, nuclear power stood at 1% of supply. Iran's record for seasonal peak electricity demand is approaching 56,000 Megawatts (MW)[524], or 56 Gigawatts (GW) in summer.

Figure 3: Iran Electricity Generation by Fuel

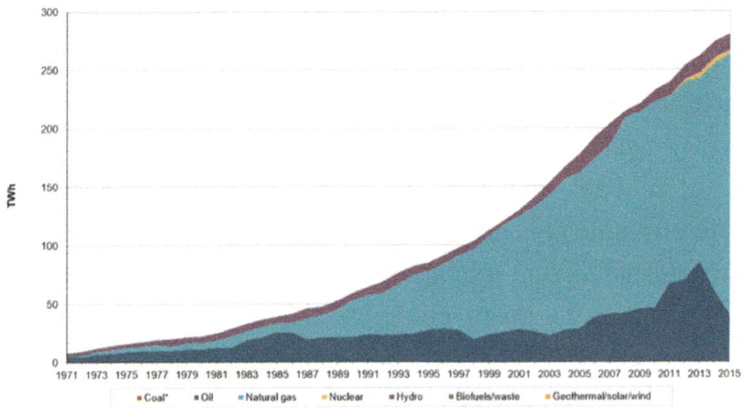

Source: International Energy Agency

The chart above also indicates the remarkably rapid growth rate of electricity generation, and thus demand, in Iran prior to the imposition of the most restrictive UN oil sanctions. From 1995-2012 demand grew by a compound average rate of over 6.5% per year,

524. https://financialtribune.com/articles/energy/67589/iran-electricity-consumption-near-all-time-high

while from 2012-15 growth slowed below 2% per year, based on data reported to the International Energy Agency (IEA).[525] The growth of natural gas consumption other than for electricity generation during these periods, at 8% and 3.8% per year, respectively, is similar enough to the pace of electricity demand growth to suggest that Iran's economy was not electrifying more rapidly than the overall growth of domestic energy consumption in the economy in the last two decades. If anything, electrification may have slowed relative to total energy demand while the tightest sanctions were in place and the economy contracted.

(It is also evident that during the period of strongest sanctions, a pronounced shift toward oil-burning power plants occurred, and away from gas. This might be explained by depressed internal oil prices, relative to gas, if fuels were priced at market rates in Iran. However, Iran's fuels markets have been heavily distorted by subsidies of various kinds for many years. It seems equally plausible to conclude that a shift in the domestic energy mix was required to provide an outlet for oil—or its byproducts—that had to be produced due to reservoir integrity concerns or other factors, but could not practically be stored until sanctions would end at some future date.

The implementation of the JCPOA and the end of most nuclear-related sanctions, along with subsequent improvements in GDP growth, do not necessarily signify a return to the previous long-term trajectory of electricity growth. Post-JCPOA analysis of Iran's future energy supply and demand by the Stanford-Iran Project[526] arrived at an implied rate of forecasted electricity demand growth of 2-2.5% through 2040. This was the based on a sector-by-sector buildup of demand, including much more rapid growth in the

---

525.http://www.iea.org/statistics/statisticssearch/report/?year=2015&country=IRAN&product=ElectricityandHeat
526. https://iranian-studies.stanford.edu/sites/default/files/publications/the_outlook_for_natural_gas_electricity_and_renewable_energy_in_iran_2.pdf

energy-intensive industrial sector. That is broadly consistent with an earlier forecast by BMI Research, which saw Iran expanding its power generation by 2.7% per year from 2016-23.

The Iran Project's report also found significant scope for efficiency improvements in Iran's existing gas-fired power plant fleet, through a combination of upgrading simple-cycle gas turbine plants to combined-cycle operation, and through replacement of gas-fired steam power plants by new combined-cycle plants. These upgrades would result in higher net generation of electricity, as well as significant fuel savings. When added to the expected increases in Iran's natural gas production over this period, the study found ample supplies of gas to sustain or even increase the high share of gas in power generation in Iran, depending on the level of future natural gas exports.

Recent reports indicate[527] that Iran is continuing its legislated program of privatizing state-owned power generation assets. The Thermal Power Plants Holding Company will apparently sell another 11,570 MW of generation to private companies, in keeping with plans to reach private ownership of 80% of the power sector, by capacity.

III. Resource Endowment and Availability

Continued expansion of Iran's gas-fired power generation capacity from its already substantial installed base requires long-term access to reliable sources of natural gas on a commensurate scale. Unlike countries that must rely on imported gas from distant producers, transported either by pipeline or as liquefied natural gas (LNG) via ocean-going tankers, for Iran this consideration is reduced to a

---

527. https://financialtribune.com/articles/energy/81327/iran-govt-divesting-stake-in-thermal-power-plants

simple question of the size and accuracy of its domestic natural gas reserves.

Iran is a substantial net exporter of energy[528] but most of that is in the form of crude oil and petroleum products. Although its state-owned energy industry aspires[529] to become a significant exporter of LNG in the future, current gas exports are modest, consisting of pipeline deliveries to neighboring Turkey, Armenia and Azerbaijan. The latter volumes are largely offset by imported gas from Turkmenistan.

Despite low levels of gas trade, since 2013 Iran's reported "proved reserves" of natural gas have ranked as the largest in the world, following a substantial downward revision[530] in Russian gas reserves. The new figures put Iran ahead of Russia and the US, even though both these countries produce a multiple of Iran's annual gas output.

As defined, proved gas reserves include only identified resources that are economic to produce[531] at current prices, using current practices and technology. Changes in price or improvements in technology can result in large shifts in proved reserves, such as the 45% increase in U.S. natural gas reserves from 2006 to 2106, resulting from improved capabilities to extract shale gas and other "tight" gas resources economically on a large scale. (*BP 2017 Statistical Energy Report*)

One measure of the sufficiency of a country's natural gas reserves is implied reserve life. This metric does not consider future field

---

528.http://www.iea.org/statistics/statisticssearch/report/?country=IRAN&product=balances&year=2015

529. https://www.wsj.com/articles/iran-seeks-rapid-reboot-for-natural-gas-exports-1453821547

530. https://www.reuters.com/article/bp-reserves/update-3-bp-cuts-global-gas-reserves-estimate-mostly-for-russia-idUSL5N0EO1I720130612

531. https://www.eia.gov/todayinenergy/detail.php?id=17151

production rates or logistics, but simply divides reserves by annual production (R/P). Iran's gas reserve life stands at 166 years at current production. By comparison, Russia's reserve life equates to 56 years at current production, while that of the US, now the world's largest producer, is just under 12 years. (*BP 2017*)

That comparison suggests that, barring any obstacles to obtaining the necessary investment or access to technology, the scale of Iran's gas reserves should provide significant headroom to expand output without running into resource constraints. A 2015 forecast[532] released by the National Iranian Gas Company indicated that with all economic constraints removed, and with large investments in new gas fields, Iran could increase its gas production to 450 billion cubic meters (BCM) per year by 2022, more than double reported 2016 production of 202 BCM. (*BP 2017*)

Iran's long-term gas potential ultimately hinges on the reliability of its published reserves. A 2005 study[533] by the Society of Petroleum Engineers (SPE) found good agreement in the treatment of the three categories of resources traditionally included in "proved reserves"--"proved developed", "proved non-producing", and "proved undeveloped"—among the agencies responsible for reserves accounting standards in the US, Canada, UK, Norway, Russia and China. All three categories represent gas that is both economically and technically recoverable[534] with additional investment. As long as Iran adheres to similar standards in its self-reported reserves, then those figures should be comparable to--and as reliable as--those of other major producers.

The largest area of potential uncertainty in this regard is in the category of "proved undeveloped" (PUD) reserves. In Iran's case,

---

532. https://pgjonline.com/2015/08/12/forecasting-irans-natural-gas-production-consumption/
533. http://www.spe.org/industry/docs/OGR_Mapping.pdf
534. https://www.eia.gov/todayinenergy/detail.php?id=17151

those account for at least 85%[535] of stated gas reserves. In other words, most of Iran's future natural gas appears to reside in fields for which a discovery well has presumably been drilled, but that may not have been delineated through additional drilling, nor have they produced a cubic meter of commercial gas. That is not necessarily a serious cause for doubt, but it is indicative of the relative immaturity of Iran's gas sector, when compared to its oil industry, which has been producing since 1908 and is now rather mature.

If anything, Iran's reliance on PUD gas reserves suggests there may be further upside in its estimates, since reserves customarily grow as fields are developed through further drilling. Historical data on U.S. natural gas provide a good example of how reserves and eventual production can multiply over time through development. Based on data from the U.S. Energy Information Administration (EIA) cumulative U.S. dry natural gas production[536] (excluding gas liquids and reinjection) totaled 1,200 trillion cubic feet (TCF)—nearly 35 trillion cubic meters, or roughly Iran's current gas reserves—even though U.S. proved natural gas reserves have never exceeded 368 TCF[537] and were for many years estimated to be below 200 TCF. Improved technology and higher prices were also important factors in that reserve expansion, and both could play a role in the future growth of Iran's reserves.

Another indication that Iran has substantial gas reserves available to develop is that 47% of current reserves[538] are attributable to the enormous offshore South Pars Field, which extends across Iran's maritime border with Qatar, where it is called the North Field. Together South Pars and the North Field form the world's largest

---

535. https://www.eia.gov/analysis/requests/ngexports_iran/pdf/full.pdf
536. https://www.eia.gov/dnav/ng/hist/n9070us2a.htm
537. https://www.eia.gov/dnav/ng/hist/rngr11nus_1a.htm
538. https://www.eia.gov/analysis/requests/ngexports_iran/pdf/full.pdf

non-associated natural gas field, i.e. gas not produced in conjunction with producing crude oil.

Qatar's production from this structure currently exceeds Iran's. In 2017 Qatar ended a self-imposed moratorium[539] on further development of the field and is in the early stages[540] of a significant expansion of output. For its part, Iran signed an agreement[541] with French energy company Total, which will take a majority stake in the next stage of its South Pars development for an investment starting at $1 billion.

That announcement, the first such deal following the signing of the JCPOA, meets an important need. At this point the most important constraints on realizing future production growth from Iran's gas reserves are financial. Production technology, geology and geography all appear to be in hand.

Even without much foreign investment, Iran has managed to expand its gas production at a 6% compound average growth rate since 2006 (EIA data through 2014), in line with the rise of domestic demand for the product. That result has of necessity been achieved with local talent and technology, due to the direct and indirect impact of sanctions.

It is also worth noting that Iran's oil and gas production are still essentially all from conventional reservoirs. Shale gas and oil resources have been discovered[542] in several parts of the country, but particularly for gas its development is likely to remain uneconomic

---

539. https://www.haaretz.com/middle-east-news/the-qatar-iran-gas-field-behind-the-diplomatic-war-in-the-middle-east-1.5480343
540. http://www.gulf-times.com/story/576882/Qatar-s-North-Field-expansion-moving-full-steam-ah
541. http://money.cnn.com/2017/07/03/news/iran-total-sign-2-billion-gas-deal/index.html
542. https://oilprice.com/Energy/Crude-Oil/Why-Irans-Shale-Oil-Discovery-Wont-Add-To-The-Glut.html

for many years. The focus of the U.S. oil and gas industry on shale gas in the previous decade, and on shale or "tight oil" in the current decade, was largely a function of the maturity and decline of most conventional U.S. oil and gas fields. Iran still has a full menu of conventional gas discoveries to exploit at lower cost.

Finally, in addition to large hydrocarbon reserves, Iran's territory includes significant renewable energy resources. Their current contribution and future potential will be explored in the section below.

IV. Iran's Alternative Energy Options

At one time "alternative energy" encompassed a wide array of possible energy sources, including some that were far from commercial availability or competitive cost. And since "alternative" mainly conveyed choices to alleviate what was then regarded as an excessive reliance on expensive and presumably scarce oil and gas, even coal qualified in certain forms. Today, with coal under intense environmental pressure and nuclear power in either economic or regulatory retreat in much of the developed world, alternative energy and renewable energy have become more or less synonymous as the former term falls into disuse.

Hydropower still accounts for by far the largest renewable contribution to Iran's electricity generation. Dams and other hydroelectric installations made up roughly 17% of Iran's electric utility generating capacity in 2015.[543] However, due to factors

543. http://data.un.org/Data.aspx?d=EDATA&f=cmID%3aEC

including persistent drought,[544] hydro supplied only 5% of Iran's electricity[545] that year.

These facilities exist at various scales, but more than 90% are larger than 10 MW.[546] Their output has been highly variable[547] over multi-year periods, but the trend has been generally downward, with hydropower in 2015 generating a quarter less annually than in 2007. For this reason, further meaningful expansion of hydropower in Iran appears unlikely.

By contrast, so-called non-hydro renewables, mainly wind and solar power, accounted for much less than 1% of both electrical capacity and generation in Iran as of the latest available statistics (2015), despite both potential and plans for substantial growth.

Iran's Sixth Five-Year Plan targets the addition of 5,000 MW of renewable[548] energy by 2021 out of a total of 25 GW of generation of all types to be added.[549] This appears to be based on previously announced targets from 2014,[550] anticipating 5,000 MW of renewables by 2018. This appears to be focused mainly on wind and solar power with, with 500 MW of the latter as an initial target.

It is unclear whether any of these targets can be met within the planned timeframe. However, even if they were, this would very

---

544. http://www.mei.edu/content/article/iran%E2%80%99s-renewable-energy-potential
545.http://www.iea.org/statistics/statisticssearch/report/?year=2015&country=IRAN&product=ElectricityandHeat
546.http://www.irena.org/DocumentDownloads/Publications/IRENA_Renewable_Energy_Statistics_2017.pdf
547.http://www.irena.org/DocumentDownloads/Publications/IRENA_Renewable_Energy_Statistics_2017.pdf
548. https://cleantechnica.com/2017/02/08/iran-approves-3-billion-worth-foreign-renewable-energy-investments/
549. http://www.cms-lawnow.com/ealerts/2017/04/iran-approves-the-sixth-development-plan-to-boost-investment
550. https://www.pv-magazine.com/2014/05/08/iran-reveals-5-gw-plans-for-solar-and-wind_100015040/

likely fall short of the goal of getting 10% of Iran's electricity from all renewable sources, including hydro, by 2021. If total generation grows from 2015 levels at the 2.7% per year rate cited above, the expected yield from another 5,000 MW of wind and solar installations might push renewables to 8% of all power generated, assuming no further deterioration of hydro output since 2015.

At least $3 billion of foreign investment[551] has reportedly already been committed to renewable energy projects in Iran, equivalent to only around 1,500 MW of combined new wind and solar capacity. However, both that figure and the five times larger target[552] set for 2030 represent just a small fraction of Iran's renewable energy potential, which has been estimated at 100,000 MW (100GW)[553] for wind alone, roughly equivalent to China's 2014 wind power capacity. In this respect the development of Iran's renewable energy resources other than flowing rivers is even less mature than its exploitation of its enormous natural gas reserves.

Iran's location and climate render it especially suitable for generating electricity from sunlight. Iran's renewable energy agency, SUNA, reports average annual solar intensity[554] ranging from 4.5-5.5 peak-sun-hours per day (equivalent to 4.5-5.5 kWh/m2/d). This is an indication of excellent solar resources, with the high end of that range approaching some of the most attractive locations for solar power in the U.S. southwest.

The higher the peak-sun-hours at a given location, the more electricity can be generated annually from the same quantity of solar photovoltaic (PV) panels or from a concentrating solar power (CSP)

---

551. https://cleantechnica.com/2017/02/08/iran-approves-3-billion-worth-foreign-renewable-energy-investments/
552. https://cleantechnica.com/2017/02/08/iran-approves-3-billion-worth-foreign-renewable-energy-investments/
553. http://archiv.windenergietage.de/WT24/24WT12_F9_1010_Iran.pdf
554.https://www.solarwirtschaft.de/fileadmin/media/pdf/AA_Report_BSW_Iran.pdf

facility. Thus a solar installation in southern Iran might generate, on average, over 70% more electricity[555] yearly than one in southern Germany, with much less monthly variability.

The latter point is especially important for integrating grid-connected renewables without stressing the existing power grid. Wind and solar energy are intermittent power sources, varying across daily, seasonal and annual scales. The lower the variation in peak sunlight, the more solar generation a power grid can integrate without requiring significant quantities of storage or other fast-reacting sources.

Intermittent renewables such as wind and solar power have useful synergies with natural gas, even as they compete for demand at the margin. Iran's current heavy reliance on natural gas for electricity generation is an advantage in this regard. Simple-cycle gas turbines in peak-power service can adjust quickly for fluctuations in supply due to changes in wind and cloud cover. The country's newer combined cycle power plants should have similar capabilities if they were designed to provide short "ramp" times, a measure of how quickly their output can increase or decrease.

Large-scale baseload generators, including nuclear power plants, typically cannot respond as quickly to such changes, if they can respond at all. This has become an important concern in markets with a high penetration of intermittent renewables, such as California, where the daily summer peak of solar generation is growing large enough to displace baseload generation,[556] unless the renewable "overgeneration" is "curtailed", or taken off-line.

To put Iran's renewable energy targets and investments in perspective, its neighbor across the Persian Gulf, Saudi Arabia, has

555. http://solarelectricityhandbook.com/solar-irradiance.html
556. https://www.nrel.gov/docs/fy16osti/65023.pdf

its own goal of achieving 10% of its generation from renewables[557] by 2023, but starting from nearly zero without the equivalent of Iran's head-start in hydropower. The Kingdom plans to invest up to $7 billion in solar and wind installations in 2018 alone. Similarly aggressive developments are under way among the countries of the Gulf Cooperation Council. A single solar project in Dubai, the Mohammed bin Rashid Al Maktoum Solar Park,[558] is intended to grow in phases from an initial scale of 13 MW, completed in 2013, to 5,000 MW by 2030—matching the ambition of Iran's entire renewable energy target.

A key driver of the wave of renewable energy installations across the Middle East and elsewhere is the rapidly falling cost of the associated technologies, particularly solar PV. This will be addressed in Section VII below.

V. Reducing Greenhouse Gas Emissions

Based on 2015 emissions of 630 million metric tons,[559] Iran ranks eighth among countries for carbon dioxide (CO_2), the main greenhouse gas implicated in climate change, below Canada but above South Korea and Saudi Arabia. It ranks fifth on national emissions per unit of GDP, below China and Russia but above relatively carbon-intensive Canada and Australia.

Iran is a signatory to the Paris Agreement,[560] joining most other countries of the world in the voluntary commitment to reduce its

---

557. http://www.straitstimes.com/world/middle-east/from-oil-to-solar-energy-saudi-arabia-plots-a-shift-to-renewables
558. https://www.dubaisolarshow.com/views/key-solar-projects-programmes-dubai.aspx
559. http://edgar.jrc.ec.europa.eu/news_docs/jrc-2016-trends-in-global-co2-emissions-2016-report-103425.pdf
560. http://unfccc.int/paris_agreement/items/9444.php

emissions of CO_2 and other greenhouse gases (GHGs). As of this writing, Iran has not yet ratified the agreement.

In preparation for the late-2015 Paris climate conference of the United Nations Framework Convention on Climate Change, Iran submitted its Intended Nationally Determined Commitment (INDC),[561] spelling out its planned levels of emissions reduction and the conditions attached to them. In this document, Iran's government indicated it would reduce emissions unconditionally by 4% by 2030, relative to a "business-as-usual scenario." This relatively modest level of reductions would be achieved by reducing emissions from its electric utility sector, reducing natural gas flaring, improving energy efficiency, diversifying its economy, and participating in national and international "market-based mechanisms". That term usually refers to emissions trading and other means of shifting emissions reductions between companies, economic sectors, or governments.

Iran also committed to a higher level of emissions reductions, up to an additional 8% of baseline emissions, conditioned on receiving international assistance, including technology transfer and capacity building. The estimated cost of the promised 4-12% emissions reductions was $18-53 billion.[562]

In the introduction of its INDC, the Islamic Republic referenced the country's need for continued hydrocarbon development to meet social and economic goals. Many of the specific technologies it identified as candidates for transfer relate to mitigating the greenhouse gas impact of oil and gas development, as well as in power generation, including natural gas combined cycle (NGCC)

---

561. http://www4.unfccc.int/Submissions/INDC/Published Documents/Iran/1/INDC Iran Final Text.pdf
562. http://www4.unfccc.int/Submissions/INDC/Published Documents/Iran/1/INDC Iran Final Text.pdf

and combined-heat-and-power (CHP) generation, nuclear power and renewables.

Since the electricity sector is a primary contributor to the country's current emissions, its modernization should align with Iran's emission-reduction goals. In fact, all of the main strategies for updating generation in Iran—replacing gas-fired steam plants with NGCCs, deploying wind and solar power, and building nuclear power plants—contribute to meaningful emissions reductions. However, the extent and effective cost of the resulting emissions reductions varies significantly.

One particularly useful strategy not mentioned or implied in Iran's INDC is the displacement of oil from power generation. As noted in section II above, 14% of the electricity generated in Iran in 2015 was derived from oil, mainly high sulfur fuel oil and other petroleum products. Unlike in Saudi Arabia, little or none of the petroleum used by the power sector is in the form of crude oil. This is an important distinction when it comes to considering the alternate disposition of the petroleum that might be freed up by fuel switching or alternative generation.

The direct substitution of natural gas for residual fuel oil in even an inefficient steam power plant should result in a 29% drop in CO_2 emissions per Megawatt-hour (MWh), based on the relative emission factors[563] of the two fuels when combusted. Replacing the steam plant with a modern NGCC could result in CO_2 reductions of at least 46%, due to lower heat rate[564] of the gas turbine, compared to a boiler-driven steam turbine, resulting in less fuel used for the same output.

---

563. https://www.epa.gov/sites/production/files/2015-07/documents/emission-factors_2014.pdf
564. https://www.eia.gov/electricity/annual/html/epa_08_02.html

This is analogous to the emissions reductions achieved in the U.S. utility sector as lower-priced natural gas has displaced electricity generated from coal-fired power plants. U.S. CO2 emissions from electricity generation have fallen by roughly one-fifth since 2008[565] as coal's share of the electricity market fell from 48% to 34% in 2015, while the share of generation from gas increased from 21% to 32%.

Figure 4 shows that even larger emission reductions can be achieved by displacing the output of steam generation, which has characteristics similar to coal, with renewables like wind and solar power. Iran's existing hydropower facilities also qualify as low-emission on this scale, as does Iran's sole nuclear power plant. However, nuclear power is hardly the only tool with which Iran could reduce its GHG emissions, nor, as we will see shortly, is it the least expensive.

Figure 4: Lifecycle GHG Emissions in Power Generation

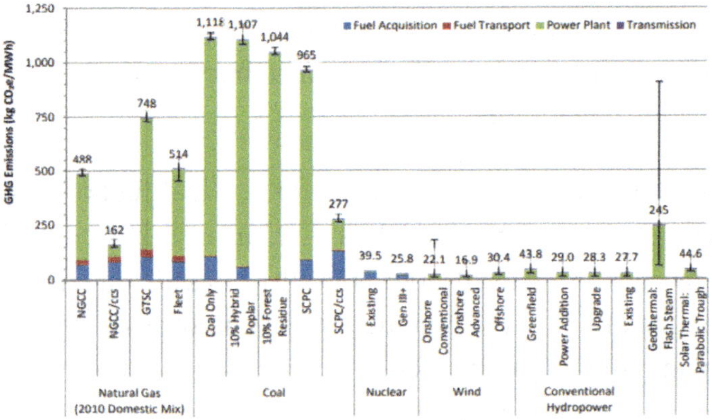

Source: DOE National Energy Technology Laboratory

The chart also illustrates the emission-reduction potential of pairing gas power generation with carbon capture and sequestration, denoted as NGCC/ccs. CCS installations capture and concentrate

565. https://www.eia.gov/todayinenergy/detail.php?id=26232

CO_2 from the power plant exhaust for injection into a long-term storage[566] site, utilization in industrial processes, or for enhanced oil recovery as demonstrated by the PetraNova project in Texas.[567] CCS on natural gas is a simpler proposition than for coal-fired power plants, with their inherent complexities of solids handling. Chevron's Gorgon liquefied natural gas (LNG) plant[568] in Australia captures and stores over 3.4 million tonnes per year of CO_2 from the natural gas feed to the plant, equivalent to the emissions from a 1,500 MW NGCC[569] plant. Similar facilities have operated in the North Sea since 1996.[570]

Iran's power sector consumed 11.6 million tonnes of fuel oil[571] in 2015—equivalent to around 200,000 barrels per day—resulting in approximately 37 million tonnes per year of CO_2 emissions. If that figure could be cut in half by shifting demand away from Iran's oil-burning steam plants and onto new, efficient NGCC power plants, it would go a long way to meeting the country's 4% emissions reduction pledge under the Paris Agreement. Eliminating oil power generation by substituting an equivalent capacity of wind or solar power would exceed that pledge.

Ending Iran's inefficient oil-based power generation would provide other significant benefits, especially in reducing local pollution. For Iran today, air quality is a more acute problem than climate change.

---

566. https://www.theguardian.com/environment/2014/oct/01/canada-switches-on-worlds-first-carbon-capture-power-plant
567. http://www.powermag.com/capturing-carbon-and-seizing-innovation-petra-nova-is-powers-plant-of-the-year/
568. http://www.zeroco2.no/projects/gorgon
569. https://blogs.scientificamerican.com/plugged-in/running-the-numbers-on-epae28099s-new-co2-regulations-combined-cycle-stacks-up-well/
570. https://www.globalccsinstitute.com/projects/sleipner%C2%A0co2-storage-project
571.http://www.iea.org/statistics/statisticssearch/report/?year=2015&country=IRAN&product=Balances

Four of the 10 worst cities[572] is a more acute problem than climate change. [573] in the world for air pollution are in Iran, and despite clean-up efforts, Teheran's air is still labeled as "unhealthy" a third of the year. A 2004 paper[574] published in the Iranian Journal of Health Science and Engineering found that merely substituting natural gas for fuel oil virtually eliminated the SOx emissions of four of the country's power plants, while reducing smog-forming NOx by 45-80%. New NGCC power plants have very low emissions of such local pollutants.

However, even if Iran were inclined to take this step, it must secure the additional gas supply, requiring additional upstream gas development, as well as an outlet for the displaced oil. As discussed in the next section, the latter concern is far from trivial, even for an oil-exporting nation like Iran.

VI. The Export Factor

Prior to the imposition of the UN sanctions that were lifted after the signing of the JCPOA, Iran had the second-highest oil exports within the Organization of Oil Exporting Countries (OPEC), shipping over 2.5 million[575] barrels per day (MBD) as recently as 2010. Its recovery, post-sanctions, has been gradual, and conditions had changed in the interim. Iran's exports reached 2.1 MBD in 2017, [576] but that would only rank fourth, at best, within OPEC,[577] since

572. http://www.al-monitor.com/pulse/originals/2016/11/tehran-air-pollution-alarming-levels-schools-shut.html
573. http://www.bmj.com/content/348/bmj.g1586
574. http://www.bioline.org.br/pdf?se04008
575. https://asb.opec.org/index.php/interactive-charts/oil-trade
576. https://oilprice.com/Latest-Energy-News/World-News/Iran-Oil-Exports-Hit-777-Million-Barrels-In-2017.html
577.http://www.opec.org/opec_web/static_files_project/media/downloads/publications/ASB2017_13062017.pdf

the exports of both Iraq and the United Arab Emirates have grown in the meantime.

Moreover, when sanctions ended, global oil markets were experiencing a severe slump caused by the rapid growth of U.S. shale oil production. Iran's resurgent exports compounded this situation until OPEC joined with Russia and other non-OPEC[578] producing countries to restrain output and attempt to rebalance the global market and thereby shrink the significant bulge in global oil inventories that had accumulated. Following the lifting of a decades-old restriction limiting exports of all but a select few U.S. oil grades, the U.S. has become a significant oil exporter[579] in its own right, adding another new factor to the global market.

With the oil market so delicately balanced, at least for the moment, and OPEC's members bound by their agreement with Russia, Iran would not necessarily be free to export additional oil liberated from power generation by a shift to other fuels or energy sources, even though the quantities are equivalent to roughly 10% of Iran's current oil exports. The fact that the fuel burned in Iran's steam plants is not crude oil, but rather a grade of residual fuel oil, complicates matters further.

Under a new regulation adopted in 2016,[580] the International Maritime Organization (IMO) will restrict the global use in ocean-going vessels of high-sulfur fuel oil of the kind produced by Iran's refineries, starting on January 1, 2020. The regulation cuts allowable sulfur content of the fuel from the current 3.5% to 0.5%, requiring very significant changes in refinery operations in order to meet these

578. https://www.ft.com/content/4cd8dce2-beec-11e6-9bca-2b93a6856354
579. http://www.washingtonexaminer.com/oil-exports-climbing-two-years-after-40-year-old-ban-lifted/article/2638865
580. http://www.imo.org/en/MediaCentre/PressBriefings/Pages/MEPC-70-2020sulphur.aspx

new specifications or supply the marine market with other types of fuel that already conform.

Few other outlets exist for such high-sulfur fuels, beyond continuing to burn them domestically. As described in a 2016 report[581] by Platts, the global refining industry has several options for complying with the IMO's new rule, including the diversion of higher-quality fuel from other markets, shifting refinery inputs to higher quality crude oil, and investment in new process equipment to substantially reduce the sulfur content of this material or convert it into other, more acceptable products. At this point the response appears to be a work in progress,[582] with several of the above solutions having lead times that exceed the time remaining before implementation.

So in order for Iran to displace enough fuel oil from power generation to make a material difference in its CO2 emissions, or to create a new high-value product stream for domestic consumption or export, its refineries would have to install sufficient upgrading capacity to process these volumes, at a cost of billions of dollars. The potential returns on such an investment might be attractive, at least initially, but it would compete with desired investments in oil and gas development and production facilities. The alternative would be to attempt to sell this material at distressed prices to refiners elsewhere that already have such capacity.

Thus, in Iran's case, the rationale for justifying investments in renewable energy or nuclear power on the basis of freeing-up oil currently used in power generation for export at a favorable economic uplift looks highly questionable. Under the circumstances, it appears that at least until significant investments have been made in Iran's oil refineries, new generation from either

581.https://www.platts.com/IM.Platts.Content/InsightAnalysis/IndustrySolutionPapers/SR-IMO-2020-Global-sulfur-cap-102016.pdf

582. https://www.businesswire.com/news/home/20170828005198/en/New-Low-Sulfur-Requirements-Marine-Bunker-Fuels-Causing

renewable or nuclear sources would instead compete directly with gas-fired power generation, and ultimately with the economics of future gas development and exports.

Iran is already a modest net exporter of gas. As noted in section III above, all of Iran's present gas exports occur via pipeline. Any significant increase in exports would likely have to be in the form of LNG, requiring investment in liquefaction facilities, in tandem with investment in additional production to provide the large, stable supply of natural gas necessary to support the multi-decade economics of LNG development.

Several LNG projects have been proposed for Iran but some have subsequently been withdrawn, including one with India's state oil and gas company,[583] ONGC. Russian company Gazprom reportedly signed[584] an agreement for LNG with Iran in late 2017. Other possibilities such as floating LNG facilities[585] have also been raised, yet all of these potential projects face significant uncertainties as the global market absorbs the output of new LNG facilities in Australia and the onset of LNG exports backed by U.S. shale gas. Significantly, French oil major Total's agreement[586] to invest in the development of the South Pars gas field is intended to supply Iran's *domestic* market.

Fundamentally, Iran has gas to export, whether from reserves still in the ground or liberated from current power generation needs. For more than a decade the key to large-scale gas development—and hence exports—by Iran has been access to the necessary technology and finance, and by far the largest impediment to both has been Iran's nuclear program and the suspicions that have surrounded it.

583. http://www.thehindu.com/todays-paper/tp-business/ovl-drops-plan-to-build-iran-lng-facility/article22853465.ece
584. https://www.lngworldnews.com/gazprom-in-iran-lng-cooperation-deal/
585. http://www.lngworldshipping.com/news/view,iran-ponders-lng_49826.htm
586. https://www.ft.com/content/3b530f02-5f1d-11e7-8814-0ac7eb84e5f1

In principle, the JCPOA addressed those. However, doubts about the sustainability of the agreement, particularly with regard to continued U.S. participation under the current administration,[587] leave a cloud over the attractiveness of international investment in Iran's energy sector.

Nor are these investment needs small. In August 2017 Iran's oil minister Zanganeh cited a figure of $200 billion[588] over five years to revitalize Iran's oil and gas industry and capitalize on its opportunities. Up to three-fourths of that should come from foreign investment. In this context recent deals with companies like Total represent only a down payment.

Gathering the required sums could prove challenging. The global oil and gas industry has more potential opportunities today than it appears to have the appetite to pursue. A bruising oil price slump from 2014-16 caused most international oil companies to reduce their capital budgets. The rapid post-slump recovery of U.S. shale drilling is a factor, as well, moderating near-term prices while offering quicker returns on investment, compared to most large conventional oil and gas projects, such as those on offer in Iran. The result is something of a buyer's market in "upstream" opportunities. That may explain why BP, the 5th largest international oil company by market capitalization, has opted out[589] of investment in Iran at this point.

---

587. https://www.cnn.com/2018/01/12/politics/president-donald-trump-iran-deal-waiver/index.html
588. http://www.iran-daily.com/News/199102.html?catid=3&title=Foreign-funds-for-Iran-s-oil-sector-a-top-priority--Zanganeh
589. https://www.ft.com/content/b73c0e9e-ce7b-11e6-b8ce-b9c03770f8b1

VII. Comparing the Costs of Iran's Electricity Options

The global energy industry is experiencing change on an enormous scale, particularly when it comes to the shifting future mix of generation that will meet growing demand for electricity. The most recent forecasts from organizations like the International Energy Agency and U.S. Energy Information Administration have increased the share likely to be taken by wind, solar and other renewable energy sources, along with slower but steady growth for natural gas, while moderating their view of the future market share of nuclear power. (See Figure 5.)

Figure 5: Renewables and Gas Lead World Electrical Capacity Growth

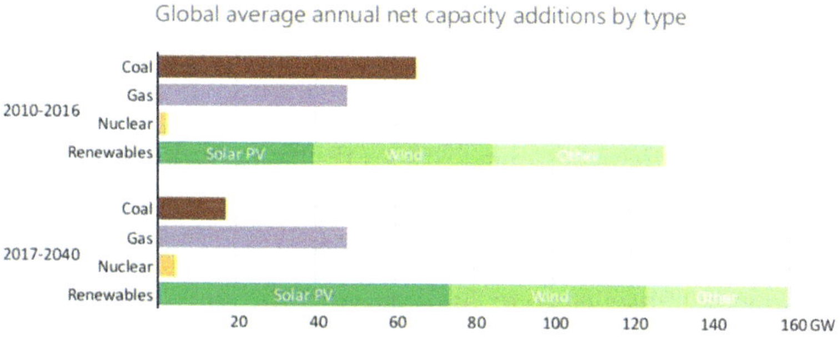

Source: IEA World Energy Outlook 2017

These shifts reflect sharp changes in conventional wisdom since the middle of the last decade. Notions of "peak oil" and its corollary, "peak gas," have wilted under a wave of production from conventional and non-conventional oil and gas resources, depressing prices and instilling a new sense of abundance. Meanwhile, renewables seemed likely to remain expensive well into the future have become competitive with conventional power plants, at least where the resources they tap are plentiful.

By contrast, after the recent cancellation of a flagship nuclear power project in South Caroline, the expected U.S. "nuclear renaissance" has dwindled to a pair of reactors under construction in Georgia. Projects in France, Finland and the UK have experienced severe cost and schedule over-runs, while Germany is abandoning nuclear power even as its policy of "Energiewende"[590]—literally "energy turn"— seeks to reduce GHG emissions by increasing the share of wind and solar power.

The result of these developments has been a surprising rearrangement of the relative cost of the main electric generation options. The "gold standard" for comparing the economics of these technologies is the "levelized cost of energy", or LCOE. It takes into account the differences among power sources with low capital but high lifetime fuel costs, and others with high capital costs but low or even essentially zero fuel expenses, along with differences in the timing of construction or installation.

Figure x below depicts ranges of LCOEs in U.S. dollars per MWh for a broad array of generating technologies on a consistent set of assumptions concerning fuel prices, interest rates, and policies—in this case in the absence of technology-specific subsidies such as the tax credits or "feed-in tariffs" that renewables have enjoyed in a number of markets. The analysis was performed by Lazard in 2016. The range bars for solar PV, gas combined cycle, and nuclear power have been highlighted for clarity.

---

590. http://www.dw.com/en/germanys-nuclear-phase-out-explained/a-39171204

Figure 6:

Unsubsidized Levelized Cost of Energy Comparison

Certain Alternative Energy generation technologies are cost-competitive with conventional generation technologies under some scenarios; such observation does not take into account potential social and environmental externalities (e.g., social costs of distributed generation, environmental consequences of certain conventional generation technologies, etc.), reliability or intermittency-related considerations (e.g., transmission and back-up generation costs associated with certain Alternative Energy technologies)

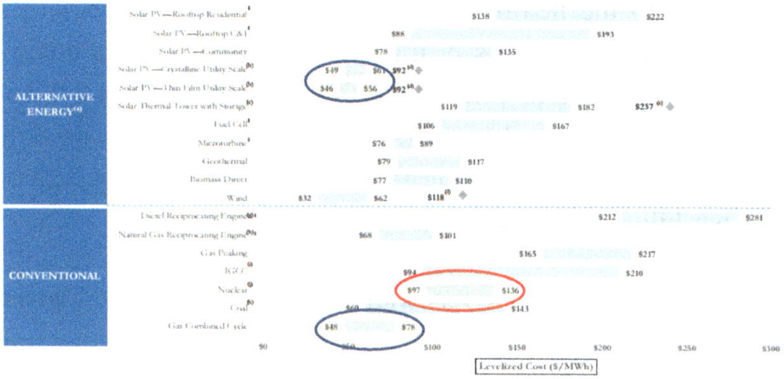

Source: Lazard "Levelized Cost of Energy Analysis - Version 10.0"

As shown here both NGCC (gas combined cycle) and utility-scale solar power appear to be considerably cheaper than nuclear power in the absence of subsidies, even without considering local factors like Iran's abundant sunshine and effective natural gas prices much lower than the U.S. prices on which this analysis was based.

The analysis is also sensitive to the cost of capital. With inflation as high[591] as in Iran, and high interest rates, capital-intensive projects that require investment over a period of years, such as nuclear power plants, would see their LCOE shift to the right (higher). That would accentuate the economic advantage of solar arrays that are built in a factory and can be installed onsite fairly quickly.

Recent solar power bids in Saudi Arabia, coming in below the LCOE range at an effective price less than $20/MWh,[592] suggest how much solar prices have fallen just in the last two years. It also highlights

591. https://tradingeconomics.com/iran/forecast
592. C:\Users\gsws\Documents\My Work\Consulting\. https:\www.fool.com\investing\2017\10\10\middle-easts-low-solar-prices-are-a-game-changer.aspx

the advantage of regions with very high annual "insolation" or solar intensity

In addition to the helpful decline in the cost of solar power, the deployment of renewables in Iran benefits from policies instituted by the government to promote it, directly and indirectly. These include a five-year feed-in-tariff (FIT),[593] paying solar developers an above-market rate for their electricity production, equivalent to $0.15 per kilowatt-hour (kWh).

Another policy advancing renewables less directly is the legislated privatization[594] of Iran's state-owned generators. The government aims to increase private ownership of the power sector from the current 60% to 80%, while increasing annual renewable energy installation to 1,000 MW. With few of Iran's utility customer paying enough for electricity to cover the cost of generating it, renewables could become the most profitable generating assets, as long as the FIT remains in place.

VIII. Conclusions

Iran's energy economy is fueled by a few crucial inputs and driven to meet a relatively small set of needs. On the supply side of this ledger, the hydrocarbon resources within its territory are abundant and, in the case of natural gas, unsurpassed. On the demand side, the country's economy requires rising energy production to provide export revenue, sustain GDP growth, and meet the needs of consumers. Increasingly, the energy supplied within the country

593. https://www.pv-magazine.com/2014/05/08/iran-reveals-5-gw-plans-for-solar-and-wind_100015040/
594. https://financialtribune.com/articles/energy/81327/iran-govt-divesting-stake-in-thermal-power-plants

must be cleaner, both to reduce endemic urban air pollution and to meet the commitments Iran made under the Paris climate agreement.

However, when it comes to the other crucial input to the energy sector, the investment needed to develop Iran's resources and convert them into usable domestic energy or exportable goods, the industry is capital-constrained. That is not unusual when the government owns the oil and gas reserves, pipelines, refineries, and export facilities, as well as deciding how the significant cash flows generated by oil exports will be spent. Funds that in a publicly traded oil company would be reinvested in future production are often diverted to other purposes for which the government needs cash.

This explains why, despite external oil sales of around $36 billion in 2016,[595] Iran must seek foreign investors to participate in the continued exploitation of its share of the massive South Pars gas field and the revitalization of its oil fields. It also explains why the sanctions imposed by the UN and U.S. were effective in coercing Iran into the negotiations that led to the Joint Comprehensive Plan of Action on its nuclear program.

If the goal of Iran's energy sector is to produce as much energy as affordably possible for the growth of its domestic economy and energy exports, then the country's government appears to have better, more cost-effective options available than to spend scarce capital on a robust build-out of nuclear power plants and fuel enrichment.

There is every indication that renewable energy and modern gas-fired power plants can supply the foreseeable electricity needs of Iran's domestic economy, while reducing emissions of both local pollutants and greenhouse gases like CO_2. And because these

595. https://www.statista.com/statistics/223231/opec-net-oil-export-revenue-streams-by-country/

technologies are cheaper in terms of both up-front investment and the levelized cost of their energy output, compared to nuclear power, this would leave more money to be invested in developing oil and gas fields for export purposes, compared to the convoluted and costly path of using nuclear power to displace for export some of the oil and gas currently burned there for electricity.

Iran has ample resources of natural gas and renewable energy—wind and sunlight—with which to grow its energy economy, while its nuclear ambitions have functioned as a large and sustained impediment to obtaining the foreign direct investment needed to undertake such growth. At least from an energy perspective, the continued pursuit of the government's nuclear program can only be described as an enigma.

APPENDIX

Title V of the Nuclear Nonproliferation Act of 1978

United States Assistance to Developing Countries

Section 501. Policy and Report

The United States shall endeavor to cooperate with other nations, international institutions, and private organizations in establishing programs to assist in the development of non-nuclear energy resources, to cooperate with both developing and industrialized nations in protecting the international environment from contamination arising from both nuclear and non-nuclear energy activities, and shall seek to cooperate with and aid developing countries in meeting their energy needs through the development of such resources and the application of non-nuclear technologies consistent with the economic factors, the material resources of those countries, and environmental protection. The United States shall additionally seek to encourage other industrialized nations and groups of nations to make commitments for similar cooperation and aid to developing countries. The President shall report annually to Congress on the level of other nations' and groups of nations' commitments under such program and the relation of any such commitments to United States efforts under this title. In cooperating with and providing such assistance to developing countries, the United States shall give priority to parties to the Treaty.

Section 502. Programs

(a) The United States shall initiate a program, consistent with the aims to cooperate with developing countries for the purpose of—

(1) Meeting the energy needs required for the development of such countries

(2) Reducing the dependence of such countries on petroleum fuels, with emphasis given to utilizing solar and other renewable energy resources

(3) Expanding the energy alternatives available to such countries

(b) Such program shall include cooperation in evaluating the energy alternatives of developing countries, facilitating international trade in energy commodities, developing energy resources, and applying suitable energy technologies. The program shall include both general and country specific energy assessments and cooperative projects in resource exploration and production, training, research and development.

(c) As an integral part of such program, the Department of Energy, under the general policy guidance of the Department of State and in cooperation with the Agency for International Development and other Federal agencies as appropriate, shall initiate, as soon as practicable, a program for the exchange of United States scientists, technicians, and energy experts with those of developing countries to implement the purposes of this Act.

(d) For the purposes of carrying out this Act, there is authorized to be appropriated such sums as are contained in annual authorization Acts for the Department of Energy, including such sums which have been authorized for such purposes under previous legislation.

(e) Under the direction of the President, the Secretary of State shall ensure the coordination of the activities authorized by this title with other related activities of the United States conducted abroad, including the programs authorized by the Foreign Assistance Act of 1961.

Section 503. Report

Not later than twelve months after the date of enactment of this Act, the President shall report to the U.S. Congress on the feasibility of expanding the cooperative activities established into an international cooperative effort to include a scientific peace corps designed to encourage large numbers of technically trained volunteers to live and work in developing countries for varying periods of time for the purpose of engaging in projects to aid in meeting the energy needs of such countries through the search for and utilization of indigenous energy resources and the application of suitable technology, including the widespread utilization of renewable and unconventional energy technologies. Such report shall also include a discussion of other mechanisms to conduct a coordinated international effort to develop, demonstrate, and encourage the utilization of such technologies in developing countries

About the Authors

Dr. Ali Ahmad is a faculty member at the American University of Beirut (AUB) where he leads the Energy Policy and Security in the Middle East Program at AUB's Issam Fares Institute for Public Policy and International Affairs. Dr. Ahmad's research and lectures cover nuclear security, energy policy, and economics. Prior to AUB, Dr. Ahmad spent two years as a Research Fellow at Princeton University's Science and Global Security Program where he worked on informing nuclear diplomacy with Iran and studied the prospects of nuclear energy in the Middle East. During the fellowship, Dr. Ahmad worked with Princeton colleagues to develop solutions for the technical challenges in nuclear negotiations with Iran. Dr. Ahmad's work on energy and the Middle East has so far resulted in several journal articles and book chapters, one book (forthcoming), and several op-eds and interviews with regional and international media. Dr. Ahmad holds a degree in physics from the Lebanese University and a Ph.D. in Energy Engineering from Cambridge University. Outside of academics, Dr. Ahmad advises governments and investment firms on a variety of energy issues.

Dr. Patrick Clawson is Morningstar Senior Fellow and Director of Research at The Washington Institute, where he directs the Iran Security Initiative. He has authored more than 150 articles about the Middle East and international economics and is the author or editor of eighteen books or studies on Iran. Dr. Clawson appears frequently on television and radio, and has published op-ed articles in major newspapers including *The New York Times*, *The Wall Street Journal,* and *Washington Post*. He has also testified before congressional committees more than twenty times and has served as an expert witness in more than thirty federal cases against Iran. Prior to joining The Washington Institute, he was a senior research professor at the National Defense University's Institute for National Strategic Studies, a senior economist at the International Monetary Fund and the World Bank, and a research scholar at the Foreign Policy Research Institute.

Jessica Drun is a fellow at the Party Watch Initiative. Previously, she was a project associate at the National Bureau of Asian Research, where she managed and assisted with the organization's Taiwan programming, as well as its annual People's Liberation Army conference. Her research interests include cross-Strait relations, Taiwan domestic politics, US-China relations, and US-Taiwan relations. Previously, Ms. Drun held positions at the National Defense University, the Center for Strategic and International Studies, and the Project 2049 Institute. Ms. Drun has published in the Jamestown Foundation's *China Brief*, *War on the Rocks*, *The Global Taiwan Brief*, and *The Diplomat*, as well as been quoted in *The New York Times*, *The China Post*, and *Quartz*. She graduated with an M.A. in Asian Studies from Georgetown University's School of Foreign Service in 2015 and an A.B. in International Affairs from the University of Georgia in 2011. She is fluent in Mandarin Chinese and conversational in Min Nan Chinese.

Dr. David von Hippel is a Nautilus Institute Senior Associate based in Eugene, Oregon. His work with Nautilus has centered on energy and environmental issues in Asia, and particularly in Northeast Asia. He has done extensive analyses of the patterns of fuels use and prospects for energy efficiency and energy sector redevelopment in North Korea, trained and worked with a group of Northeast Asian energy researchers to develop and evaluate the energy security implications of different energy paths for their countries, researched and assembled a regional workshop on the environmental impacts of power grid integration in Northeast Asia, evaluated the prospects for "clean coal" technologies in China, prepared reviews of rural electrification options and of the impacts of climate change/sea-level rise in Asia and the Pacific, and contributed to many other Nautilus projects. In addition to his work with Nautilus, he works and has worked for a number of private and public agencies, including the World Bank, the United Nations and Tellus

Institute/the Stockholm Environment Institute-US (Boston, MA) on a wide variety of topics ranging from utility planning and energy efficiency in the Middle East to reducing greenhouse gas emissions in several states and the Midwest region in the United States. Dr. Von Hippel holds M.S. and Ph.D. degrees in Energy and Resources from the University of California-Berkeley, and M.A. and B.S. degrees from the University of Oregon.

Dr. Robert Ichord Jr. is a nonresident senior fellow at the Atlantic Council's Global Energy Center. He comes to the Atlantic Council after almost forty years of U.S. Government service in advancing U.S. international energy interests. He retired in January 2016 from the U.S. Department of State, where he served as deputy assistant secretary for energy transformation in the Energy Resources Bureau from 2011 to 2015. He recently established Ichord Ventures, LLC, which provides energy analysis and consulting services to both the private and public sectors. At the U.S. State Department, he helped launch the new Energy Resources Bureau established by Secretary Clinton and worked to further U.S. foreign policy interests in energy security, sustainable development, and global climate change. Prior to working for the U.S. State Department, Dr. Ichord worked for the U.S. Agency for International Development (USAID) in the Europe and Eurasia Bureau as Chief of Energy and Infrastructure, as well as in the Asia and Near East Bureaus. Dr. Ichord is a graduate of Denison University (B.A.), The Fletcher School of Law and Diplomacy (M.A.L.D.), and the University of Hawaii (Ph.D.), where he was awarded a fellowship from the East West Technology and Development Institute and completed a dissertation on Petroleum Development in Southeast Asia. He is the recipient of the George Marshall and the Science and Technology awards from USAID and received a Superior Honor Award from the Department of State for his work with Europe in forging the Athens Energy Treaty creating the Energy Community in the Balkans. At the Atlantic Council, he is leading an initiative on Transforming the Power Sector in Developing Countries.

Alexa Lee serves as an independent energy researcher and she currently works for a technology trade association as Environment Manager for the Asia-Pacific region. The research and perspectives she is presenting are her own. At the trade association, Ms. Lee coordinates information and advocacy efforts related to tech companies' energy and environmental policies in China and across Asia. Ms. Lee received a Prudential Fellowship at the Brookings Institute's Center for East Asia Studies, where she researched and organized policy discussions on China, cross-strait relations, and the impacts of the Trans-Pacific Partnership on the Asian economy. She interned at the U.S. House of Representatives Foreign Affairs Committee and spent a year at the U.S. Chamber of Commerce where she assisted advocacy dialogue on Asia trade and business discussions. Ms. Lee had participated in the APEC Women Empowerment Initiative on transportation policy where she conducted ground research in Indonesia and Vietnam. She holds a master's degree in International Economics from Johns Hopkins University School of Advanced International Studies. Ms. Lee is a native of Taiwan and she speaks Mandarin, Japanese, and Bahasa Indonesian.

Robin Mills is the CEO of Qamar Energy, based in Dubai. Mr. Mills is an expert on Middle East energy strategy and economics. He is the author of two books, *The Myth of the Oil Crisis* and *Capturing Carbon*, columnist on energy and environmental issues at Bloomberg and *The National*, and comments widely on energy issues in the media, including the *Financial Times*, *Foreign Policy*, *The Atlantic*, CNN, BBC, Sky News and others. He worked for a decade at Shell, concentrating on new business development in the Middle East, followed by six years with Dubai Holding and the Emirates National Oil Company, and spent two years as non-resident fellow for energy at the Brookings Doha Center. He is a fellow of the Columbia SIPA Center on Global Energy Policy and a senior fellow at the Iraq Energy Institute. He holds a first-class degree in Geology from the University of Cambridge, and speaks five languages including Farsi and Arabic.

Henry Sokolski is executive director of the Nonproliferation Policy Education Center and teaches graduate classes on nuclear policy at the University of Utah and the Institute of World Politics. He was recently appointed Senior Fellow for Nuclear Security Studies at the University of California at San Diego. Previously, he worked in the Pentagon as Deputy for Nonproliferation Policy, as a consultant to the National Intelligence Council, as a member of the Central Intelligence Agency's Senior Advisory Group, as a Senate military legislative aide to a member of the Senate Armed Services Committee, and as a special assistant on nuclear energy to the chairman of Senate's Tennessee Valley Authority oversight committee. He served on two congressional commissions on the prevention of WMD proliferation and has authored and edited numerous volumes on strategic weapons proliferation issues, including *Best of Intentions: America's Campaign against Strategic Weapons Proliferation* and *Underestimated: Our Not So Peaceful Nuclear Future.* Mr. Sokolski earned his graduate degree in political science from the University of Chicago.

Geoffrey Styles is Managing Director of GSW Strategy Group, LLC, an energy and environmental strategy consulting firm. Since 2002, he has served as a consultant and advisor to organizations and executives, particularly through the application of scenario planning. His industry experience includes 22 years at Texaco, Inc., culminating in a senior position on Texaco's leadership team for strategy development, focused on the global refining, marketing, transportation and alternative energy businesses, and issues such as climate change and sustainable development. Previously he held senior positions in alliance management, planning, commodity trading, supply and distribution, and risk management. Mr. Styles also served on NASA's Senior Management Oversight Committee for Space Solar Power. He earned an M.B.A. from Berkeley's Haas School of Business and a B.S. in Chemical Engineering from University of California, Davis. His "Energy Outlook" blog has been quoted by *The Wall Street Journal* and other media and was named one of the "Top 50 Eco Blogs" by the *Times of London*.

Stephen Thomas is a researcher in energy policy with more than 30 years of experience. His work is international in scope and the main areas of research are on economics and policy towards nuclear power, liberalization and privatization of the electricity and gas industries, and trade policy on network energy industries. In recent years, his research has included: an evaluation of the economics of nuclear power, an assessment of the impact of the marketization of essential services on low-income consumers, an assessment of the European Commission's Directives aimed at opening the electricity sector in Europe to competition, an assessment of the South African government's attempts to develop a new design of nuclear power plant, a review of the corporate policies of the major European energy utilities, and a review of methods of funding nuclear power decommissioning liabilities.

Leonard Weiss is a visiting scholar at Stanford University's Center for International Security and Cooperation and a member of the National Advisory Board of the Center for Arms Control and Nonproliferation in Washington, DC. For more than 20 years, he worked in the U.S. Senate as Sen. John Glenn's staff director on the Governmental Affairs Committee, writing legislation and leading investigations in the areas of nuclear proliferation, nuclear safety, and government organization and management. He was the chief architect of the Nuclear Nonproliferation Act of 1978 and legislation creating the Defense Nuclear Facilities Safety Board. He is a former tenured Professor of Applied Mathematics and Engineering at Brown University and the University of Maryland.

Made in the USA
Monee, IL
03 November 2026

67383141R00207